BE YOUR OWN CONTRACTOR!

BUILDING YOUR DREAM HOUSE
CREATING NEW SPACES
MANAGING THE PROJECT YOURSELF OR
USING A GENERAL CONTRACTOR

JAMES M. SHEPHERD

Dearborn
Financial Publishing, Inc.

Publisher: Kathleen A. Welton
Acquisitions Editor: Patrick J. Hogan
Associate Editor: Karen A. Christensen
Senior Project Editor: Jack L. Kiburz
Interior Design: Lucy Jenkins
Cover Design: Design Alliance, Inc.

Published by Dearborn Financial Publishing, Inc.

Printed in the United States of America

93 94 95 10 9 8 7 6 5 4 3

Library of Congress Cataloging-in-Publication Data

Shepherd, James M.
Be your own contractor! / James M. Shepherd.
p. cm.
Includes index.
ISBN 0-79310-496-3 (pbk.) : $24.95
1. House construction. 2. Contractors. I. Title.
TH4811.S47 1992
690′.837—dc20

92-29670
CIP

Table of Contents

List of Figures

Preface

I have written *Be Your Own Contractor!* to teach you how to manage the building of your home and save money, time and energy.

This book contains a wealth of knowledge useful in building a home that I have acquired in my ten years as a general contractor. In addition, I have spent the last several years in evaluating research to keep up with the latest and best methods of home building, including consultation with many experts in various fields. I have concentrated on providing practical, up-to-date, energy-saving principles of house construction.

To be successful as a contractor, it is vital that you have a sound basis for your construction. The guidance in this book will help you make the informed decisions necessary for quality specifications and plans. It describes the materials among which you must choose in order to balance your desire for the best against the limits imposed by your pocketbook.

This book does not show you how to lay bricks or how to frame a house. It is intended, rather, that the owner will subcontract all the labor to trained and experienced trade contractors while serving as the overseer or general contractor.

If you prefer to work through a professional general contractor, this book will be most useful in helping you understand what you're going to get in the completed house. Misunderstanding is usually caused by incomplete specifications prepared by the general contractor, lack of knowledge on the part of the owner as to what the specifications mean or both.

If your goal is to expand the living space of an already-built house, this book also discusses the special techniques for a job of this type.

Others who can benefit from this book:

- Investors who rehabilitate homes, so they can easily spot homes with strong profit potential after improvements and manage contractors once the project has begun.
- REALTORS®, so that they can recognize the quality of the home they are selling and be able to offer sound advice about its best features and how to correct the poor ones.
- Sellers of building supplies, so that they can give sound advice on which product is the best for each stage of construction.
- Bank officials who are engaged in monitoring construction loans, so they can inspect the progress of construction with confidence that the job is being done properly.
- Young contractors just starting in the business, who will find the information in this book very helpful and very difficult to locate in any other single source.

Introduction

This book is organized as follows:

- Chapters 1 and 2 cover the areas of house plans, specifications, building contracts, the bidding process, financing the construction and insurance. These are the tasks you must accomplish before actual construction begins.
- Chapters 3 through 21 discuss the building process in the order in which the steps occur. Each chapter covers how to cost the labor and material of that phase of construction and how to manage it if you are the general contractor. In addition, extensive information is provided in each chapter to assist the homeowner in making decisions on how to build and equip his or her house.
- Chapter 22 discusses how to expand an already-built house. It includes the various choices for expanding living space and special construction techniques used in expansions that are not necessary in new-house construction.
- The appendixes contain detailed information of construction specifications, complete costing tables, details on how to lay out the batter board system, a construction schedule and a listing of contractor's tasks indicating which type of subcontractor does which jobs. Of special interest to those living in very cold areas is Appendix F, "The Super Insulated House."

The level of detail in the book is based on the need of those who will be their own general contractor. Even those who will work through a professional general contractor need to have a working knowledge of the house building process so they will be able to follow this process with some assurance that the house is being built according to the terms of the contract.

Planning To Save Money

Achieving savings in building a home starts with proper costing. Each chapter includes a section describing how to compute the costs of that housing component; this allows you to determine the total cost of the house before any construction begins. Appendix B illustrates the costing process.

There are two ways to reduce the cost of your house. One is to use cheaper materials and the other is to build less. The use of cheaper materials is the least desirable of the two alternatives. The end product will deteriorate faster, require more maintenance and give you less comfort and satisfaction than a house built with quality materials. By cutting corners with "bargain" materials of any kind, you are inviting problems that may be very expensive to correct. For example, poor framing lumber will cause warping in the floors, ceilings and walls. Poor-quality windows and doors will result in higher fuel costs.

The much better choice is to build less space. Choose a smaller house or postpone certain aspects of the construction that can be finished later when your financial position improves. You can leave part of the house unfinished or plan on expanding in the future, including in your present plan the basics that will readily accommodate the expansion. For example, pour the footings for a planned later expansion along with the footings for the current house. Then cover the footings with soil and grass until you are ready to use them. This choice also reduces labor cost.

Managing Construction

If you are going to be your own general contractor, you must know how to check that the work is being done properly and to schedule each step at the proper pace. Each chapter contains suggestions on how to avoid problems by thorough and detailed planning and thus prevent delays in construction and potential increases in costs.

On the other hand, do not overmanage. Every trade contractor must be allowed to go ahead with the work he or she has contracted to perform without undue interference from the owner/general contractor. Naturally, if the contractor is not performing the job according to the plans and specifications, it is your responsibility to step in and get the job straightened out. Be careful, however, that you are not guilty of taking the time of the trade contractor and crew unnecessarily by a failure to have material on site in the proper quantities or by an overabundance of questions to satisfy your curiosity. It is always a good practice to work with the contractor rather than with a crew member.

Another caution is that before scheduling a trade contractor to begin a job, ensure that all of the necessary preliminary work by others has been completed. Failure to check this measure in many cases will justify extra payment to the trade contractor for lost time.

Appendix D presents a detailed schedule for the entire construction task. It indicates the sequence of the various jobs with explanatory notes regarding materials and other information.

Avoiding Direct Hire

If you are to be the general contractor, it is not advisable to get involved in the direct hire of labor or with payment on an hourly basis. Once you establish yourself as the employer, you will need to keep a volume of records to handle the deductions for withholding federal and state income taxes, deductions for and matching social security, worker's compensation insurance payments and other items. In addition, payment for labor on an hourly basis tends to drag out the job unless proper, close supervision is available. Your job is much easier if you work solely on a contract basis: let the trade contractors employ the labor.

Conserving Energy

Given the cost of fuel today and the indications that prices will continue to rise, the efficient application of energy-saving principles to your house construction becomes increasingly important.

Although energy is discussed throughout the book, the most significant aspects are covered in the following: Chapter 11, "Air Infiltration," Chapter 12, "Insulation," Chapter 13, "Solar Heating," Chapter 14, "Heating and Cooling" and Appendix F, "The Super Insulated House."

The objective of these chapters is to assist you in the application of energy conservation measures and the selection of the best heating and cooling systems to fit the requirements in your environment. Before you make any selections, study these chapters thoroughly.

Doing It Yourself

You may wish to do some of the work yourself for the satisfaction of having built your own house, at least in part, or to save money. Although there is nothing wrong with this approach, consider the following questions:

- Do you have the necessary skills to do a creditable job?
- Do you have the time to do the job?
- Will doing it yourself hold up other work and perhaps cost you more money in the long run?
- Will your work make it difficult to pin down responsibility should something go wrong?

If you can give yourself satisfactory answers to these questions, then by all means go ahead and do the job. Whatever your choice, this book will make the experience of building a house easier to manage.

Read the book from cover to cover to get a working knowledge of its contents. Then go back over it again at a slower pace, studying in detail those areas that pertain to building your house before you make your major decisions. This procedure can save distress and make building your own home a satisfying experience.

1

Plans and Preliminaries

The Beginning

You can begin the process of building your house in one of two ways. Either select a lot you like and then decide on a house plan that fits that particular lot, or select a house plan and pick a lot to suit it. Whichever you choose, the matching of the two is very important.

House Styles

The most common styles of houses are the rancher, two-story or colonial, story-and-a-half, split-level and split-foyer designs (see Figure 1.1).

- The *rancher* consists of only one level, although it may have a basement.
- The *two story* or *colonial* consists of two above-ground floors with each floor about the same size. It may also have a basement.
- The *story and a half* also has two floors, but the second floor is built within the roof and is thus less spacious than the first floor. Light to the second floor can be provided by separate dormer windows or a shed dormer of several windows together as one unit. The shed dormer is usually in the rear of the house.

FIGURE 1.1 House Styles

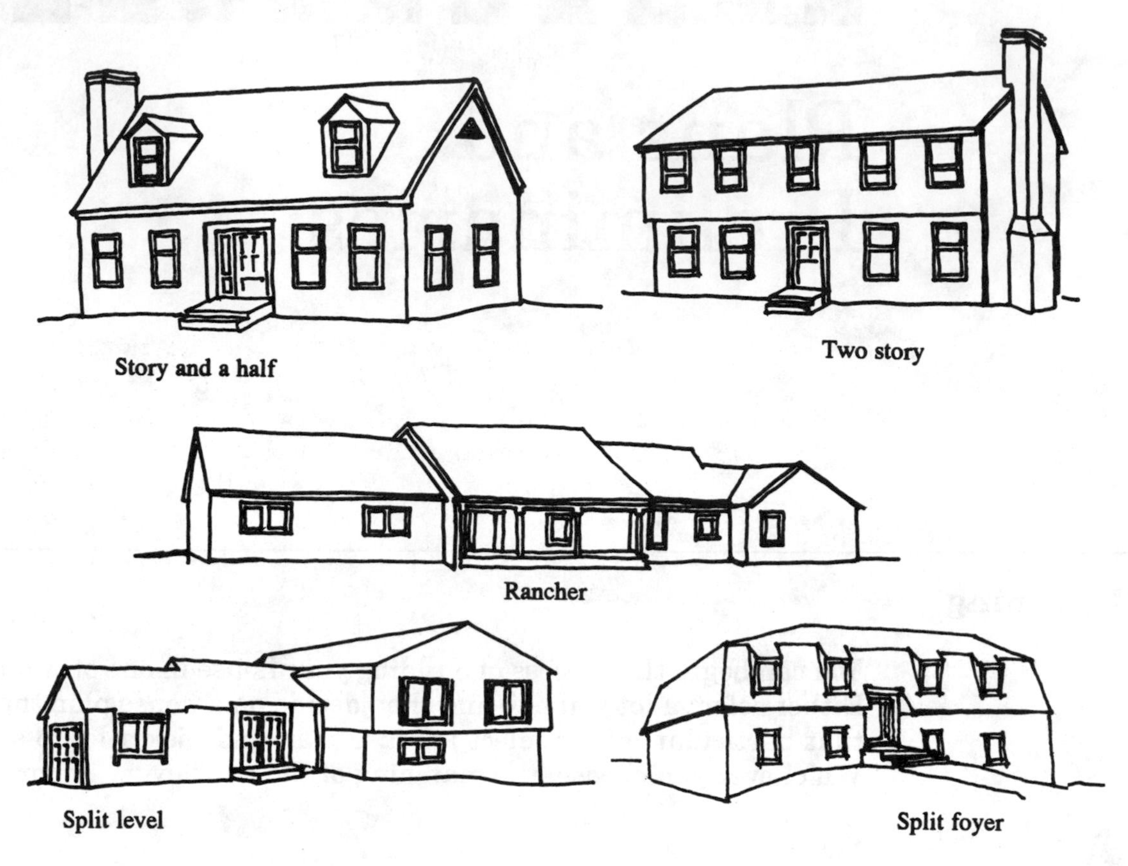

- The *split level* has three different levels of floors with intervals of about four feet between them, rather than the usual eight feet.
- The *split foyer* is similar to the two-story house except that the main entry is split between the elevations of the two floors. The bottom floor is usually partially underground.

The Right Match of House Style and Lot

Most styles of houses will fit most lots, but there are pitfalls that should be avoided. Do not match a rancher to a lot with a substantial forward slope. Ranchers look best if they appear to be hugging the ground, but a forward sloping lot exposes an exaggerated foundation in the front, which detracts from the low look of the rancher. If you have no choice, this problem can be minimized by planting the proper shrubbery. Help from a good landscaper is useful.

FIGURE 1.2 Rancher with Basement

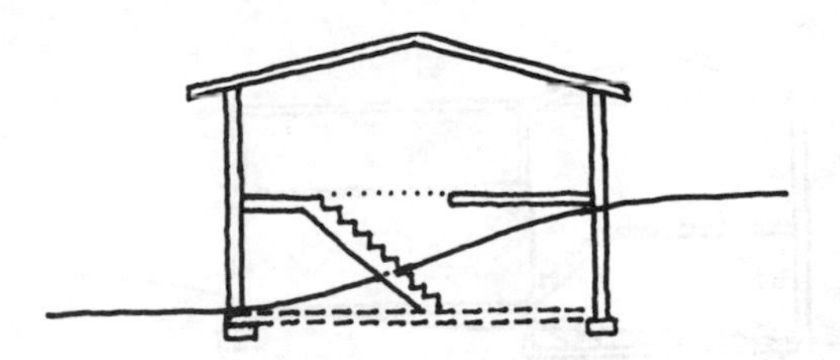

Ranchers, two-story and story-and-a-half houses are ideally suited for lots that are more or less flat. They can also be attractive on lots that slope to the sides or rear, as in Figure 1.2, particularly if the plans include a walk-out from a lower living area. This is also an economical way to increase the size of the living space, since the lower level is an extension of the foundation.

Split levels are ideally suited for side sloping lots and split foyers for rear sloping lots. (see Figure 1.3)

In matching your house to the lot, allow room for outdoor decks, patios and porches within the building setback lines. Properly placed in the plan, these elements substantially enhance the livability of a house.

Garages

Do not forget the garage. If it is attached to the house, make sure that it fits the lot without encroaching on the building setback line.

Also make certain that the garage will conform to the slope of the lot. On a side sloping lot, if you lower the garage to meet the natural lot line, you may increase the requirement for steps from the house proper to the garage floor to the extent that a section of the garage is lost because of the space required by the steps. On the other hand, if the garage is not dropped, the need for fill dirt to build up the garage floor, the driveway and portions of the lot may be substantial and may increase landscaping problems. The solution is usually a compromise.

Matching the house to the lot is very important. If you feel the need for help, by all means consult a professional architect, engineer, landscaper, contractor or other knowledgeable person. If you have to pay a fee (the normal practice), it will be money well spent.

FIGURE 1.3 Split Level and Split Foyer on Sloping Lot

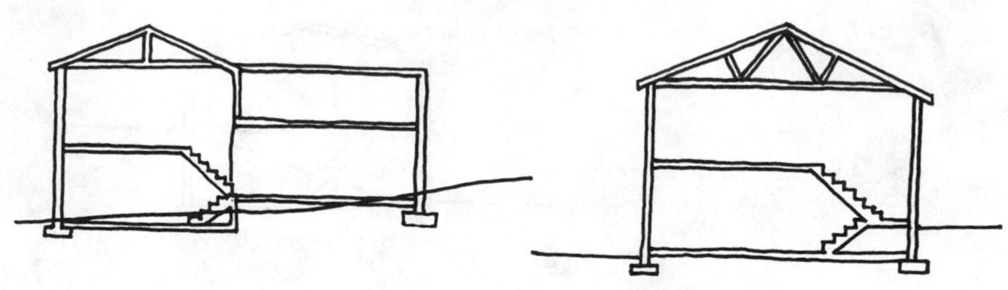

Building Systems

Three standard systems to consider in selecting the method to build your house are the modular, the panelized and the stick built.

The Modular House

Great advances have occurred in the design and construction of the modular type of house, and it has advantages that might be very important to you. It is built in sections at the factory and generally includes the complete structure with the plumbing, heating and electrical work completed and the finished walls and floor already installed. The size of the section is limited to what can be shipped over the highway, so the design of the house is somewhat restricted. Combining several sections, however, can yield some very attractive designs, including two-story houses and contemporary models with vaulted ceilings.

The package supplied by the manufacturer is nearly complete, the owner furnishing only the site work, footings, foundation, final grading and water, sewer and electrical hook-ups. Many manufacturers provide their own erection crews or will recommend local crews to do the final assembly.

The two principal advantages of the modular approach to building are the speed by which the structure can be completed and the saving in cost, since the basic components are built on the assembly line at the factory. The principal disadvantage is that choices in size, shape and exterior and architectural styles are limited by transportation factors and by what is available from the manufacturer.

The Panelized House

The panelized construction method is more flexible than the modular system in its application, because the materials are only partly assembled at the factory. The framing is usually shipped with the walls in panels of about 16 feet in width or less. Sometimes this arrangement includes the windows, doors and insulation with the siding partially installed.

Floor systems are partially assembled, and the roof framing may consist of prebuilt trusses. The remainder of the framing is often precut to some degree. The package supplied by the manufacturer may include the framing, doors, windows, shingles, electrical fixtures, plumbing fixtures, exterior trim, siding, finished hardware (medicine cabinets, doorknobs, towel and paper racks and so forth) and interior trim material. The owner must provide the site work, all masonry and concrete work, all labor to assemble the materials in the package, electrical, heating, plumbing work (less any fixtures provided in the manufacturer's package), wall finish, flooring, tile, painting, final grade and water plus sewer and electrical hook-ups.

In both modular and panelized houses, the materials and labor provided by the manufacturer vary among the different companies and should be checked carefully in determining the final cost of the entire construction job.

The advantages of the panelized house are that it provides much more flexibility in size, shape, exterior finish and architectural style than the modular home. It should save one or more weeks in building time compared with the stick-built house, and the management of the materials is simplified because so much of it is provided in the package. The panelized house also offers cost savings in that materials are bought by the manufacturer in very large quantities.

The Stick-Built House

Of the three alternative systems, stick-built construction offers the greatest flexibility in style, shape and size of the house and gives the owner the best opportunity to control the quality of the labor and materials. Because the house is built mostly from uncut lumber, it takes longer to complete and requires the owner/general contractor to spend more time and effort in working with a larger list of suppliers to ensure that the right materials are delivered to the site at the right time.

Given the greater level of supervision required for a stick-built house, the information contained in this book emphasizes that construction method.

Specialized House Designs

In recent years, several additional types of houses have become very popular: the log house, the earth sheltered house and the geodesic dome house.

The Log House

Log houses range from modest backwoods cabins to complete year-round homes for large families.

Log homes differ from those built by conventional methods in that their exterior walls are constructed of logs with diameters ranging from six to ten inches. The logs are the framing, the insulation and the inner and outer finish.

Log homes are usually constructed in one of three ways:

Traditional This system is the pioneer approach: the builder goes into the woods and cuts and trims logs or buys the unmilled logs from a supplier. This construction method requires a great deal of skill. A lot of work is involved in the on-site cutting and shaping of joints and corners. It can be very difficult, inaccurate and time-consuming. In addition, because the logs are not symmetrical, large gaps between them must be filled to make the structure weather tight. This process is not recommended for amateurs.

Round log method In this method, the builder buys logs that have already been milled to remove the bark and to shape them so that the diameter is uniform. The problem of how to fill the spaces of varying widths between the logs has been substantially reduced, but the task of cutting and shaping the joints and corners remains.

The engineered system With this method, the builder is provided the log material already cured, cut and shaped ready for installation. The work is substantially reduced, and the sealing of the house is much more effective. Figure 1.4 illustrates three of the many variations in milling logs to provide the basis for a tightly sealed house. Note that the pattern on the right has been shaped to drain the water away from the joints between the logs. In addition to the shaping, an air-tight seal is obtained by applying suitable caulking material between the logs.

There has been much discussion about the energy efficiency of the log home. During 1981–82, the National Bureau of Standards conducted energy tests of a conventionally built house and a log house located about 20 miles north of Washington, D.C.

FIGURE 1.4 Typical Log Patterns

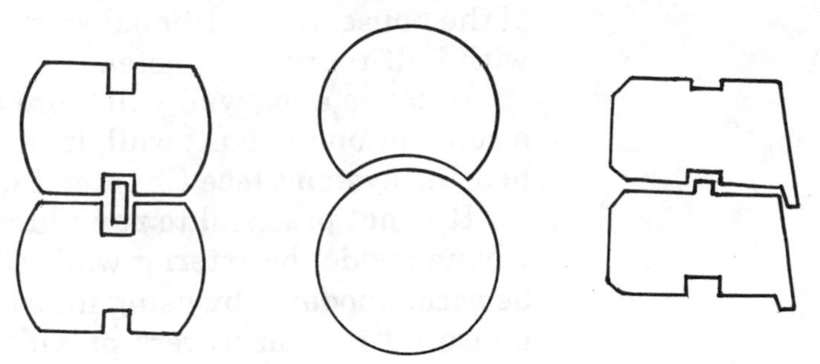

The conventionally built house had 2×4 stud walls with 3½″ of fiberglass insulation and a plastic vapor barrier. It was covered on the outside with ⅝″ thick wood siding and finished on the inside with sheet rock. The wall had an R–12 insulation rating. See Chapter 11, "Insulation."

The log house was built with exterior walls of 7″ solid square logs with an R–10 insulation rating.

Both houses were built on a concrete slab on grade with 1″ thick insulation at both inner and outer surfaces of the footing. In addition, both houses had insulation in the roof/ceiling with an R–34 rating.

Each house was equipped with a centrally located 4.1-kilowatt electric forced-air heating plant equipped with a 13,000 BTU/h split vapor-compression air conditioning system.

The results of the test were as follows:

- Both houses used the same amount of energy during the 14-week winter heating period.
- During the three-week spring heating period, the log house used 46% less energy than the conventionally built house.
- During the 11-week summer cooling period, the log house used 24% less energy then the conventional house.

The R–value of the walls for both the log house and the conventional house was too low for the test area, however. Wall R–values of R–19 are recommended for the test area; much higher values are needed for those very cold areas across the northern tier of states, Alaska and Canada.

The R–value of a log wall may be increased by installing rigid foam insulation sheets against the interior of the wall and finishing it with paneling or sheet rock. Alternatively, a 2×3 or 2×4 stud wall can be built on the inside of the wall, insulated with fiberglass and covered with a plastic vapor barrier and finished with paneling or sheet rock.

If it is important to retain the appearance of logs in the interior of the house, the additional stud wall with insulation can be covered with half-round logs instead of paneling or sheet rock.

Because a log wall will store and retain heat much better than a conventionally built wall, it can be more effective in passive solar heating systems (see Chapter 13, "Solar Heat").

It is not practical to run plumbing, heating ducts and electrical wiring inside the exterior walls of a log house. Most house plans can be accommodated by using interior walls for this purpose. Some log house kit manufacturers provide milled slots at the bottom of the inside of the exterior wall in which electrical cable can be run and covered.

Manufacturers provide a great deal of information in their brochures.

The Earth-Sheltered House

Earth-sheltered houses are structures in which large areas of the exterior walls and/or roof are in contact with masses of earth.

Figure 1.5 (designed by Survival Consultants) illustrates two methods by which this earth mass is obtained: it can be (1) piled against the exterior walls (berms) and/or (2) piled on top of the roof.

Those areas where the structure contacts earth masses must be built of masonry material strong enough to hold the weight of the soil and to compensate for the stresses the soil generates. These subgrade surfaces of the house must be effectively waterproofed and drainage systems must be installed to take away the water.

The advantages of the earth-sheltered house are:

- Large areas of the exterior of the house are not exposed to the sun and the weather and hence the exterior maintenance is greatly reduced.

FIGURE 1.5 Typical Earth-Sheltered House

- Contact with large earth masses stabilizes the temperature of the subgrade walls and roof to that of the earth's crust. Generally, the temperature of the earth's crust is lower than that of outside air in the summer and higher in winter. In addition, large masses of earth and masonry change temperatures very slowly. These features produce a structure that requires very little energy to heat and cool.
- The design of the earth-sheltered house is very suitable for passive solar heating as the major, if not the sole, supplier of heat.

The disadvantages of the earth-sheltered house are:

- Its cost may be higher than conventional construction. This higher cost, however, can be paid back in a short time by the energy saved in heating and cooling.
- The structure must be stronger than conventional construction in order to withstand the weight and stresses created by the large masses of soil.
- Repairs to waterproofing and drainage can be very costly.

Proper plans and supervision of construction are very important in building an earth-sheltered house. The designer should be experienced in this field.

The quality of waterproofing and drainage is so important that this work should be supervised or at least inspected by an experienced professional before it is covered with earth. Elastomeric roofing is an excellent material for waterproofing masonry.

Before buying plans, check with the local building codes to determine whether any provisions in the code will cause problems with your design. If so, solve these problems with the local building authorities *before* going ahead.

The Geodesic-Dome House

The geodesic-dome house, shown in Figure 1.6, is a highly specialized form of housing that continues to gain in popularity. It is the strongest of architectural forms. As the name implies, it is built more or less in the shape of a sphere. It may consist of ⅜ths of a sphere, ½ or ⅝ths. The house requires a foundation (not part of kits provided by manufacturers) and can be built with a full or partial basement. Plans are available from dome kit manufacturers for structures ranging from fewer than 1,000 square feet of living space to more than 3,000 square feet.

FIGURE 1.6 Typical Geodesic-Dome House

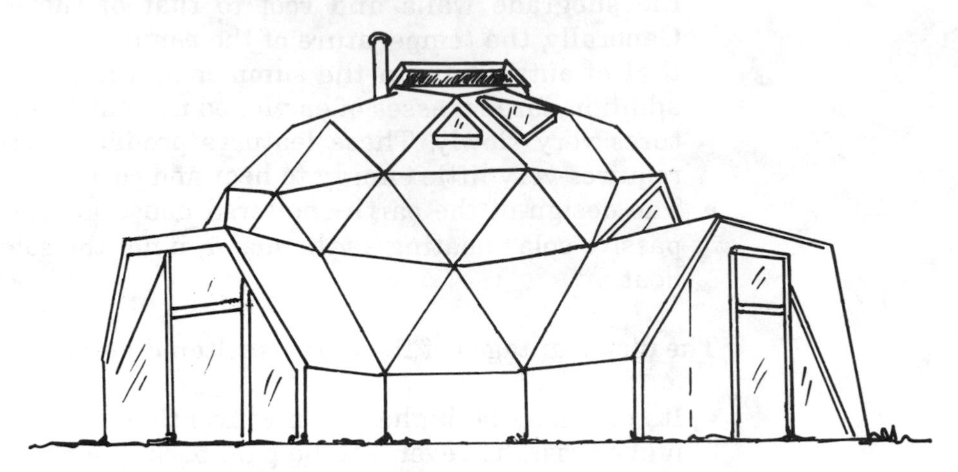

The primary design element of the dome house is the triangle, selected for its complete stability. Triangles are assembled to form the sphere that provides the roof and exterior walls of the structure.

One of two generally accepted methods can be used for designing and building the sphere. One method is the construction of a spherical frame of 2×4 or 2×6 lumber or steel framing, which is then covered with a skin; the other is to make prefabricated triangular components at a factory, which are assembled on the job to form the sphere.

Because of the extensive requirement for bevel cuts and the use of many different angles for assembly, building the dome home from scratch on the site requires the skill and experience of experts. It is far easier to build a dome home by buying a kit from a manufacturer and assembling the kit at the site.

Dome houses are quite different from conventional houses. Here are some of the characteristics:

- The interior of the structure gives the occupant the feeling of spaciousness not normally found in a conventional house.
- In ½- and ⅜-sphere domes, the vertical and horizontal curvature of the exterior walls may present a problem in the location of cabinets and furniture. This problem is more or less eliminated with the ⅝-sphere dome.
- It is not easy to install windows and doors on the curved surface of a sphere. Usually, windows and doors are specially made to accommodate the triangular design, or they are installed in dormers. Because of the structure's shape, it is often possible to light the entire dome with just a few skylights.

- **It is most important that the outer sphere be efficiently sealed and waterproofed, preferably by professionals.**
- **Changes in temperature expose the sphere to expansion and contraction, which sometimes cause creaking of the structure.**
- **Because of its shape, air inside the sphere will circulate very efficiently. For this reason, minimize the use of interior walls to enhance your heating system.**
- **Because the shape of the dome house provides the least exterior exposure, it reduces heat loss through its outer surface. Thus, domes are more energy-efficient than houses of other shapes assuming, of course, that proper sealing and insulation have been provided and that the heating system is well suited for the local climate and the house.**

Manufacturers of geodesic dome kits are quite willing to provide much more detailed information regarding dome houses in general and the kits they manufacture.

Plans

To build a house properly requires a good set of plans. You may choose from among several sources, but before you arrange for your plans, visit as many houses as is practical to get a feel for room sizes, traffic flow among the various rooms, details of finishes and other information that goes into the design of a house.

Plans from a Professional Architect

An architect will probably be the most expensive source of plans initially, but this alternative may be the least expensive in the long run, particularly since it will ensure that your house design fits your lot and obtains the maximum benefits possible from modern approaches to construction, particularly in the field of efficient energy. This subject is discussed in detail in Chapters 11, 12, 13 and 14.

A professional architect can provide a plan that closely matches your needs, especially if you have an unusual house design in mind or if you want accuracy in your selection of a particular architectural style. The architect's specifications will more precisely and thoroughly define the options of acceptable materials and construction methods and thus give you greater assurance of quality building.

Because architects often personally favor a particular style, try to find one whose style matches your tastes. Don't hesitate to ask to see some of his or her work.

Plans from a Mail-Order Service

Various mail-order planning services have become a popular alternative. Many of them do excellent work. You may not find exactly what you want, but most plans can be modified to some extent by a competent drafter. Be cautious in this regard, because what may appear to be a simple change may actually cause major structural problems. Avoid changing load-bearing walls in the foundation and framing (extending or shortening non-load-bearing walls is a fairly simple change) and avoid major changes in the roof line.

Be aware that mail-order plans do not relate to any particular lot. Be prepared, then, to have to decide on the foundation modifications to adapt the plan to your lot (especially for sloping lots). If this is necessary, get professional advice.

If you select a modular or panelized house, most manufacturers provide plans as part of the package.

Plans from Local Sources

Usually, many local sources provide plans through drafters and designers with various skills. Check with local general contractors and building inspectors for recommendations. Many of them do good work, but do some investigating before selecting one. If your local building code requires certified plans, you will have the additional task of getting this certification, which can only be provided by a professional architect or engineer.

Reading Blueprints

Reading blueprints of house design is not difficult and the homeowner/general contractor should make an effort to learn how to do so. If you are using a local architect, engineer or designer, through your contacts with them (involving many visits), you should develop a fair ability to read the prints when they have been finalized. In any case, know your plans. You will be working with them for several months and living in the results for many years. A glossary of terms relating to the house-building business is located in the back of this book.

2

Contracts, Financing and Insurance

The Building Contract

Building a new house is the most expensive project that most people undertake. It is very important, therefore, that you exercise care in developing the building contract to ensure that the house is properly built and your investment protected.

The laws and customs governing construction vary from state to state. A well-drawn contract for building in one state will not necessarily serve the owner adequately in another state. For this reason, before signing any building contract have it reviewed by a local attorney who is experienced in this type of legal work. Most professional architects or engineers are experienced in preparing building contracts.

If you are working through a general contractor, as a minimum your contract should contain the following:

1. A complete set of plans (blueprints) drawn by an architect, engineer or designer experienced in house construction and design, which should include the following elements:

   - Foundation plan
   - Floor plan for each floor
   - Elevation for each side

   Note: A plan *denotes the horizontal or bird's-eye view and an* elevation *a vertical view.*

   - Construction details of typical walls

- Construction details of special features
- Electrical plans showing, as a minimum, switches, outlet receptacles, TV outlets, central vacuum outlets and electrical fixtures, fans and heaters
- Plumbing plan showing, as a minimum, the location of all fixtures requiring water supply and/or waste piping, to include, for example, the ice maker in many refrigerators
- HVAC plan showing, as a minimum, the location of the major elements such as furnace, heat pump inside and outside element, major duct lines, air supply outlets and return air grilles and exhaust duct lines and exits through the exterior walls for exhaust fans

2. A complete set of specifications (The blueprints contain much of the construction information, but not all, by any means. For example, what kind of heating system should you have? What kind of plumbing fixtures? To spell out this information, you must have a detailed list of specifications. Appendix A illustrates a typical list of "specs" with an explanation of why each item is necessary. The remaining chapters of the book discuss choices of the various items with the advantages and disadvantages of each.)
3. A plot plan, which is a scaled drawing showing, as a minimum, the exact location of the house on the lot with distances to each lot boundary, driveway location, building setback lines and all easements. Local building officials may require the location of sewer and water lines. Check with these officials to find out exactly what is required.
4. Agreement that the general contractor will be responsible for liability insurance to protect him or her from worker's compensation acts for his or her company and covering all subcontractors.
5. Agreement as to who (owner or general contractor) is responsible for vandalism, theft of building material, damage to the house structure during the construction period and liability not covered by the general contractor's worker's compensation insurance.
6. Indication as to when construction will begin and approximately when the house will be ready for occupation by the owner. These dates are usually a guide, not specific goals. In the event, however, that the completion date is very important to the owner, a penalty clause can be included to require the general contractor to forfeit a monetary penalty for each day beyond the contractual completion date. A

penalty of this nature usually increases the contractor's bid price.

7. A statement that the house will be built in conformance with the plans and specifications.
8. A statement that changes to the contract made after signing the contract must be in writing and must include any additional cost.

If you are acting as your own general contractor and building through the use of subcontractors, the same content should be included with the following exceptions (numbers relate to those contained in the general contractor's contract above):

4. Each subcontractor must carry worker's compensation insurance and must furnish copies of the insurance certificate.
5. The owner should be responsible for carrying appropriate insurance.
6. It is highly unusual to include a time of completion penalty clause in a contract with a subcontractor, who normally has no control of the construction phasing except for his or her own part.

Remember that good plans and specifications provide a firm base for costing the house, standards by which to compare bids and a finite guide for the actual construction.

Bidding

Selecting the Bidders

Before you make the selection of contractors and suppliers to bid on your job, do some investigating:

- Suppliers can usually recommend contractors to furnish the labor for the materials the supplier sells.
- Ask for references of people for whom the contractor worked.
- Check with the local Chamber of Commerce and Better Business Bureau.
- Ask contractors of one trade the names of reliable contractors in another trade.

Handling the Bidding with General Contractors

If you are going to have your house built by a general contractor, ask for bids from at least three whom you have researched.

If you obtain only one bid, you will have no basis for examining whether or not the bid is competitive. If you get only two bids and one is high and the other much lower, you will have the dilemma of not knowing if the high bid is overpriced or the low bid underpriced.

On the other hand, three bids should provide you with a better basis for making a choice. If two bids are low and one is high, then you are probably safe in accepting one of the low bids. If two are high

FIGURE 2.1 Sample Bid Letter

1 September 1992

Built-Rite General Contracting Company
24 High Street
Center City, NC 28000

Dear Sir:

I would appreciate your bid to furnish all material and labor for the construction of my house in accordance with the plans, specifications and plot plan attached. The house is to be built on Lot 17, Scarlet Drive, Brentwood Subdivision in Center City.

I expect you to provide insurance covering all worker's compensation acts for your company and all subcontractors employed by you.

I will carry general builder's risk insurance covering vandalism, theft, fire and wind damage and general liability insurance.

If your bid is accepted, I would like to know approximately when you can begin construction and your estimate of when the house will be ready for occupancy.

If you have any questions please give me a call.

J. J. Jones
184 Central Avenue
Center City, NC 28001
Tel. (811) 234-8603

and one is low, you had better be cautious in accepting the lower bid. The low-bidding contractor may have underestimated the work and may cause problems by either cutting corners or failing to finish the contract when he or she discovers the mistake. Consider getting a bid from a fourth contractor to help solve the problem.

These bids will naturally ask for large amounts of money and should be carefully examined. Each bidder will require a complete set of plans and specifications.

Figure 2.1 is a sample bid letter to selected general contractors.

Because insurance coverage policies vary from state to state, check with your local agent for advice about the types of coverage you and the general contractor should carry.

When you receive all bids from the general contractors, compare them closely. If the contractor you accept has replied in the form of a contract to the owner, have it checked by a local lawyer familiar with this type of contract before signing. If the reply is not in contract form, have one prepared for you by a lawyer and send it to the contractor you have selected for signature. Figure 2.2 is a sample bid acceptance letter to a subcontractor and a typical response. Your acceptance letter to a general contractor should be similar.

Handling the Bidding with Trade Contractors

If you are undertaking the construction of your house as the general contractor, you will have to ask for bids from the trade contractors yourself.

Each category of work in the construction of your house should be offered to contractors and/or suppliers for bids. Get at least three bids for each item. See Appendix E for a listing of the work the trade contractors normally perform.

Some contractors prefer to furnish the material as well as the labor, and it is usually to your advantage to agree. This is particularly true for plumbers, electricians and heating contractors. It is important that you make no attempt to find "bargain" material and ask the contractor to do only the installation. By splitting the responsibility for the work between labor and material, you are creating a situation where, if anything goes wrong with the completed system, the contractor can always argue that the malfunction is caused by the equipment furnished by you, the owner, and is, therefore, not his other responsibility. Unless you are an expert, it will be difficult to prove otherwise.

Insist that bids be made in writing in sufficient detail so that you know exactly what you are getting. Give each bidder a complete set of plans and specifications and insist that the bid refer directly to

FIGURE 2.2 Sample Bid Acceptance Letter

1 September 1991

Shaw Heating and Air Conditioning Company
24 High Street
Center City, North Carolina 28000

Dear Sir:

I am pleased to accept your bid for $3,863.00 to provide all of the labor and materials for installing the HVAC system in my house at Lot #29, Jasmine Hills subdivision.

I understand that your price is based on the specifications listed in paragraph 22 including subparagraphs a through e. In addition, you will provide and pay for all permits and inspections, and would you please furnish me copies of your certificate showing that you have up to date worker's compensation insurance covering all laborers working for you on this job.

Also, to help me fit you into the schedule, fill in the blanks indicating your estimate of the time for you to perform the job and countersign the copy of this letter indicating your final agreement.

Keep the original for your records.

I am looking forward to working with you in the near future.

Yours very truly,

J. J. Jones
184 Central Avenue
Center City, North Carolina 28000
Tel: 696-0808

I verify the terms of our agreement indicated above.

I would appreciate at least   days notice for beginning the rough-in and the final. I estimate it will take me   days to complete the rough-in and days to complete the final.

A copy of my worker's compensation insurance certificate is enclosed.

Robert Shaw
Shaw Heating and Air Conditioning

those plans and specifications. Ask how much notice the contractor would like to have before beginning the work and obtain an estimate of the time it will take to complete the job. As a rule, bidders will need several days to give you their bid.

When the work is to be phased, ask for the time of completion for each phase. For instance, plumbing work is usually done in at least two phases, the rough-in and the final. These estimates will help you schedule the various parts of the whole construction project.

Requests for bids from the trade contractors are handled in a manner similar to that for the general contractors.

The Construction Loan

Financing is a complex issue that varies throughout the country, primarily because of differences in state banking laws and regulations. Policies of lenders within the same state may even vary between small towns and large cities. Visit several lenders in your area to determine the best one for you.

You will need two different types of financing: (1) is the construction or short-term loan, which is used as a source of money to pay for the house during construction, and (2) the mortgage or long-term loan, which is used to pay off the construction loan after the house has been completed.

Some lenders will issue only short-term loans, while some prefer to issue only long-term loans. Other lenders may issue a single loan to cover both the construction and the mortgage; this arrangement requires only one closing, which can reduce total closing costs.

If you are building your house by contract with a professional general contractor, your application to the lender for a construction loan is generally approved if: (1) the contractor's reputation is a good one and known by the lender, (2) your financial statement indicates that the project is within your financial means and (3) if you have at least a commitment from a lender for your mortgage if one is necessary. Get this commitment before you go shopping for the construction loan.

If you are your own general contractor, getting a construction loan may be more difficult if the lender does not know anything about your ability to manage the job. To improve your chances of getting the construction loan, consider the following suggestions:

1. **Have a complete set of plans with detailed specifications (see Appendix A) to indicate that you are completely knowledgeable about the product you are getting.**

2. Determine the total cost of the house, including breakdowns of the various parts, and back up this costing with bids from well-recognized local trade contractors.
3. Develop a detailed construction schedule for your particular job. See Appendix D, "Construction Schedule."
4. Offer collateral for the loan other than the house you are going to build, such as other real property, stocks and bonds, savings certificates or other tangible assets.

Payment during Construction

If you are building through a professional general contractor, the contract should specify in detail the method of payment. Two systems are in common use.

The first is the payment of certain sums at the completion of different stages of construction. For example, assuming that you have a $60,000 contract, the general contractor might ask for a payment of $15,000 when the subfloor has been laid (which means that much of the site work, all of the footings, all of the foundation and the floor framing have been completed); $15,000 when the framing, siding and exterior trim have been completed; $15,000 when the drywall or plaster has been completed and the final $15,000 when the house has been finished and the occupancy permit has been issued by the building inspector.

This is a simple system, but it is not necessarily an accurate one. For example, the payment of the first $15,000 may be more than the contractor has put into the house if it is a simple rancher. On the other hand, if the plan has a full or walkout basement, the $15,000 may be much less than the materials and labor put into the construction.

With the second method, the general contractor asks for a payment at the end of each month based on the status of the construction at that time. This arrangement requires a detailed billing showing the percentage of construction completed, broken down into categories such as masonry, carpentry and so forth. With this system, it is common practice for the owner or lender to retain 10% of the amount due to ensure that the contract will be completed.

Regardless of which method is used, actual payments will be monitored by the institution that loaned you the construction money. The contractor normally uses his or her own money to pay for labor and materials pending a payment by the owner or lender. The lender will have experts inspect the job each time a payment is due to determine if the amount asked for is correct.

If you are your own general contractor, you should expect to make payments somewhat differently. Methods of payment should be a part of the final contract between you and each trade contractor. Most large supply houses will bill you at the end of the month and expect payment a few days later. Some offer discounts for prompt payment, but always ask first since many suppliers do not advertise them. *Taking advantage of discounts can save several hundred dollars.*

Most of the contractors will expect payment as certain stages of construction are completed. For example, the plumbing, electrical and heating contractors usually ask for payment of about half the total after the completion of the rough-in and the remainder after the completion of the work, including the inspection by the building official's office.

Some contractors, such as masons and carpenters, may ask for payment at the end of each week based on the amount of work completed at that time. This is particularly true of small companies that need weekly payments to make their own payroll.

All of these systems are acceptable and used in the house-building business. The important point is that *whatever system you agree to, have it in writing and then abide by it.*

The Mortgage

Once the project is finished, it is common practice to pay off the construction loan with a mortgage on the house. *Before you make any contractual commitment to build a house, you should have completed the arrangements for both the construction loan and the mortgage.* A mortgage is a valid, fully enforceable contract. Be sure you understand all of the provisions of your mortgage before signing it.

When you first finance a home you are likely to make a down payment on the property, but, because you have financed the purchase, you are now in debt and the lender "owns" most of the property's value. In traditional mortgages, the monthly payments on the loan are weighted. During the first years, they are largely *interest;* in time, more of each payment is credited to the loan itself, or the *principal.* Gradually, as you pay off the principal you build up *equity,* or ownership. Your equity also increases if the value of the home increases. This process of gradually obtaining equity and reducing debt through payments of principal and interest is called *amortization.*

Until recently, most mortgages had fixed monthly payments, a fixed interest rate and full amortization (or transfer of equity) over a period of 15 or 30 years. These features worked in the borrower's favor. Inflation made your payments seem less and your property worth more, so although the payments seemed hard to meet at first, over time, they became easier.

A variety of mortgage packages in addition to the traditional fixed-rate loan are available. One of these may help you finance a home you otherwise couldn't afford, but it also may expose you to greater risks. For example, the interest rate and monthly payments may change during the term of the loan to reflect what the market will bear. Or the interest rate may fluctuate while the payments stay the same and the amount of principal paid off varies. The latter approach allows the lender to credit a greater portion of the payment to interest when rates are high. Some plans also offer below-market interest rates, but they may not help you build up equity.

In shopping for financing sources, keep in mind the following terms, which are keys to the affordability of a home:

- The *total cost minus your down payment,* or amount you finance
- The length, or *maturity* of the loan
- The size of the *monthly payments*
- The *interest rates* or rate
- Whether the payment or rates may *change* and, if so, within what limitations, if any
- Whether or not *extra payments on the principal* can be made at any time
- Whether there is an *opportunity for refinancing* the loan when it matures, if necessary

What Size Mortgage Can You Afford?

Most mortgage lenders process and underwrite loans according to generally accepted standards. The lender will want to look at the following elements of your financial situation:

- Monthly income and expenses
- Credit history
- Property appraisal
- Source of cash for down payment and settlement costs
- Employment history

These lenders will also make two calculations:

- *Housing ratio* Your total monthly mortgage payment (PITI or principal, interest, taxes and insurance) divided by your total monthly income. As a guide, this ratio should not exceed about 28%.
- *Debt ratio* The sum of your total monthly mortgage payment and other monthly debt payments divided by your monthly income. As a guide, this ratio should not exceed about 36%.

To determine the PITI, contact your bank and ask for an estimate of your monthly payment on principal and interest based on your mortgage requirements. The local tax office (city or county) can provide you the tax estimate and the insurance company can provide the insurance payment. Doing your homework before you apply for a mortgage will give you a better idea of where you stand.

Types of Mortgages

Many types of mortgages are available to the borrower. Take the time to shop around and determine which is best for you, remembering that all types have advantages and disadvantages. The more commonly used mortgages are:

Fixed-rate mortgage This type of loan is characterized by a fixed rate of interest, a long term (usually 15 or 30 years) and equal monthly installments of principal and interest until the loan is paid in full. This mortgage offers stability and the advantages of long-term payment, but interest rates may be higher than those for other types of financing. New fixed-rate mortgages may not be assumable should you decide to sell in the future.

Adjustable-rate mortgage Based on a financial index, the interest rate for an adjustable-rate mortgage will change, resulting in possible changes in the monthly payments, the term of the loan (time to pay off) and/or the principal. The financial index used may be the Federal Home Loan Bank Board's national average mortgage rate, the U.S. Treasury bill rate or the prime rate. If the financial index increases, the interest on your mortgage increases a like amount, or if the financial index decreases, the interest rate on your mortgage will also decrease. Some plans will contain caps (limits to any changes). This type of mortgage is readily available. Its starting interest rate is slightly below the market, but payments can increase sharply and frequently if the index increases. Payment caps prevent wide fluctuations in payments but can cause negative amortization. This means that your monthly payments may not be sufficient to pay even the interest. To make up for this deficit, you may lose some of your equity in the house.

FHA-insured loans Under the FHA (Federal Housing Administration) and the VA (Veterans Affairs) (below) programs, the federal government does not make mortgage loans, but rather, it insures that the borrower will repay the loan according to the contract terms. This insurance covers loans up to about $120,000, but the limit varies throughout the country. Check with your local bank for the limit in your area. The interest rate is generally lower than that for a conventional fixed-rate loan.

VA-insured loans The Veterans Administration offers a similar loan program for veterans of the armed forces only. The loan limit of this insurance is about $184,000. Like the FHA loan, the interest rate is usually lower than that for the fixed rate mortgage.

Farmers Home Administration loans This federal agency usually works through local offices to help families with relatively low incomes buy or build homes. The maximum amount of the loan varies throughout the country. Interest rates run about 2% below the rate for conventional mortgages. Contact your local Farmers Home Administration office for details.

Balloon mortgage Monthly payments are usually based on a fixed interest rate and a short term, usually five or seven years. The loan is amortized as a 30-year mortgage, but at the end of the term the entire principal is due. Therefore, the balloon mortgage offers low monthly payments but possibly no equity until the loan is fully paid. When due, the loan must be paid off or refinanced. Refinancing may be at a higher interest rate.

Graduated payment mortgage Lower monthly payments rise gradually, then level off after about five years for the duration of the term of the loan. With a flexible interest rate, additional payment changes are possible if the interest index changes. With the lower monthly payments, at least initially, it is easier to qualify, but the borrower's income must be able to keep pace with the payment increases. For the first five years, the remaining loan balance goes up slightly, instead of down. This is called negative amortization.

The mortgage market generally offers other types of mortgages as well. If your situation is unusual, consult your local banker or mortgage company to find out if an additional selection would better fit your requirements.

The Total Cost of a Mortgage

In shopping for mortgages, many people neglect to consider the total cost of the mortgage. Often this cost can be substantially reduced by a modest increase in the monthly payment:

Loan Amount	*Cost 30-Year Term*	*Add for 15-Year Term*	*Interest Saved*
$ 50,000	$ 439/month	$ 98/month	$ 61,290
$ 75,000	$ 658/month	$148/month	$ 91,935
$100,000	$ 878/month	$197/month	$122,580
$150,000	$1,317/month	$295/month	$174,870

These figures are based on an interest of 10%. The table indicates, for example, that by increasing the monthly payments on a $75,000 loan by $197 per month for a total monthly payment of $806, the mortgage term can be changed from 30 years to 15 with an interest saving of $91,935, because additional payments go directly to the principal.

Settlement (Closing) Costs

Almost every mortgage transaction involves settlement costs to cover additional expenses. A sample for a $100,000 mortgage follows:

Lender fees:	
Loan arrangement fees	$500 to $1,000
Discount points	0 to $3,000
Inspection fee (new house)	$50 to $100
Document preparation	$50 to $200
Mortgage insurance	0 to $1,000
Title charges:	
Attorney's fees	$200 to $400
Title insurance	$300 to $400
Transfer tax	$100 to $2,000
Recording tax	$200 to $400
Miscellaneous charges:	
Survey	$150 to $200
Termite inspection	$250 to $400

Prepaid expenses:	
Odd days interest	0 to $800
Homeowner's insurance	$250 to $400
Real estate taxes	$200 to $1,500
Total charges:	$2,250 to $11,800

Insurance

During Construction

The following types of insurance are necessary during the construction phase of your project:

- *Title insurance* is needed to protect your ownership of the land and the house you are building. *The amount of insurance should include the market value of both the lot and the house when completed.* Usually, the lending institution will require insurance in the amount of the mortgage only, which does not fully protect the owner.
- *Basic homeowner's insurance* covers loss from damage, fire, theft and personal coverage to those injured on your lot. In some states, an endorsement "additional extended coverage" is available for as little as $20. This endorsement dramatically broadens fire and extended coverage. In addition, special construction policies provide coverage for the dwelling and material stored on the job site.

Discuss your insurance coverage with an experienced agent. The details tend to be somewhat complicated.

- *Worker's compensation* protects you from claims by workers injured while building your house. This type of insurance is particularly important if you are acting as your own general contractor dealing directly with the subcontractors. For additional protection, hire only subcontractors who carry their own worker's compensation insurance and ask them to provide you with a certificate verifying that they do have this insurance.
- *Flood insurance* will also be required by your mortgagor in areas where a history of flooding exists. Again, make sure it covers your property fully and not just the amount of the mortgage.

After Completion of Construction

Continue all of the coverages discussed above with the exception of worker's compensation. In addition, you may want to consider *warranties* to provide coverage of house breakdowns as a whole for up to ten years. In many cases coverage is not complete. For example, if a leak develops in a water pipe in the wall, usually the material and labor costs to replace or repair the leaky pipe are covered by the warranty, but this may not be true for the material and labor costs to replace the plaster or paneling that had to be damaged in order to get to the pipe. *Learn the details of your warranty before you sign it.*

3

Starting Construction

Preconstruction Tasks

Before you begin the actual construction of your house, make sure that the following tasks are completed. The list is a guide and should be reviewed to ensure that it meets the needs for your particular project. Tasks are listed generally in the order in which they should be done.

If you are working through a professional general contractor:

- Check with the local building official to see that the house plans are approved.
- Get the approval of plans by the local subdivision building committee if required.
- Roughly stake out the house and driveway on the lot following the plot plan you have previously developed. Again make sure that:

  1. You have sufficient room for a septic system if one is required.
  2. There is enough space between the house and the boundaries of the lot to meet the local building codes.

- Complete financing arrangements.
- Have a signed contract with the general contractor you have selected to do the job.
- Write or call to thank the other bidders. This is a matter of courtesy. Many of these bidders spent a lot of time, which

usually means money, to give you the information you asked for.

In addition, if you are going to be the general contractor:

- Complete the cost of construction based on the bids you have received from the trade contractors and suppliers.
- Have a completed contract with the trade contractors you have selected to do the job.
- Write or call to thank the other bidders.
- Get the building permit. At the same time arrange for payment of any water and sewer hookup fees necessary for issue of the permit.
- Notify the electrical contractor to begin installing the temporary electrical power to the lot.
- Notify the plumbing contractor to install the temporary water hookup.

Permits

Usually, a fee is required for a building permit. It is based on the size of the house or the cost of the construction. As part of the application, you will probably be required to submit a plot plan, blueprints and specifications, which the inspection office will retain for its files. Most local building officials will review your plan and approve it or let you know if there are any code violations that will require changes.

If your plans include a subsoil disposal drain system and/or a well, you will need permits for the installation, usually issued by the local health department.

Inspections

Most local governments require inspections by them at different phases during construction. This is for your protection. Be certain that you have a list of these inspections and notify the appropriate building official when you are ready for each one. It could be very costly if you miss one and continue with the work. The building inspector could require you to rip out some of the work to expose areas that had not been inspected. If the inspector does not pass your job, see that the necessary corrections are made and a reinspection completed before you continue.

The following inspections may be required:

1. Footing and slab excavation before the concrete is poured
2. Floor framing before the subfloor is installed
3. Framing
4. Plumbing rough-in
5. Electrical rough-in
6. Heating, ventilating and air conditioning (HVAC) rough-in

The last four must be done with the interior of the framing exposed. Be sure that the drywall or plaster lathing and the insulation are not installed until completion of these inspections.

7. Plumbing final
8. Electrical final
9. HVAC final
10. Final building inspection (occupancy permit issued).

Zoning

Before you buy your lot, check with the local authorities or your real estate agent to determine whether or not the zoning of the land you are buying is suitable for the type of structure you are going to build. *In addition, check the zoning around your area; current laws should prohibit nearby construction that would detract from your house.*

Subdivision Restrictions

In some instances, the covenants of residential subdivisions require the submission of plot plans, blueprints and specifications to the subdivision building committee for its approval. The purpose of these committees is to provide economic protection for the subdivision. They want to be sure that the house you are going to build will maintain or enhance the value of the houses already built and that the architecture of the house is not so bizarre as to detract from the value of neighboring houses.

Site Work

Site work includes all of those operations that must be done before actual construction can begin. It usually includes some or all of the following:

1. On a wooded lot, clearing the trees from the construction area
2. Changing grades to fit your plan
3. Excavations for basements
4. Grading for the driveway plus filling and compacting as necessary
5. Removing excess dirt, trees, rocks and other debris
6. Installing the septic system or laying the hookup line to the public sewer
7. Drilling a well or installing the hookup line to the central water system
8. Laying out the batter-board system to locate the house

Clearing

If you have a wooded lot, remove only those trees necessary to permit construction unless you have already definite landscaping plans for the entire lot. Remember that trees can be removed quickly but it takes years of growth to replace one.

Before getting bids on the work, determine the area to be cleared for construction of the house, driveway and septic system, if one has to be installed. One of the simplest methods is to locate the corners of the house with small stakes (you don't have to be too accurate at this stage—within a foot is fine) and select an area around the house, driveway and septic system plus at least ten additional feet to allow room for the construction. Mark this area by tying tape on selected trees. Brilliant orange or red plastic tapes are excellent. This method of marking makes it easy for the site work contractor to identify the scope of the task.

House Location

House location consists of, first, the temporary location to provide the basis for lot clearing, and, after the completion of the clearing, the installation of the batter-board system, which should be accomplished by someone experienced in this work. See Appendix C, "House Layout."

Take advantage of the sun for light and heat. Face the side of the house with the most glass toward the south if your overall plan permits this arrangement. In the winter the low-hanging sun's rays will penetrate the glass area and add warmth and light inside the house. In the summer, a properly designed roof overhang can block the rays of the sun, which is higher in the sky, and thus prevent the heat from these rays adding to the air-conditioning load. (See Chapter 13, "Solar Heating.") If you are using solar heat extensively, consider the possible blockage of this type of heat by future construction on adjacent lots.

In high-wind areas, place your house to take advantage of the reduction of the wind by trees, particularly evergreens, which retain their "leaves" during the winter.

Driveway

In clearing space for the driveway, allow a wide area at the street end. If your driveway is about 10′ wide, its width at the curb should be not less than 16′ to permit cars to turn into it easily.

Culverts

Many lots require a culvert pipe to run under the driveway for proper drainage. If this culvert is located on any city/county property or right of way, check with the appropriate government office for the proper size and method of installation for the pipe. In some communities, the highway department will install the culvert pipe at no cost to the owner. If not, make certain that the installation is included in the site work contract, which should also include the establishment of the driveway base. During construction the heavy concrete and lumber trucks will pack down the base and make it stable by the time you are ready to put on the final surface.

Figure 3.1 shows three types of culvert pipes in general use.

For a finished look, the metal and nonflared concrete pipe need head walls made of brick, stone or concrete.

FIGURE 3.1 Types of Culverts

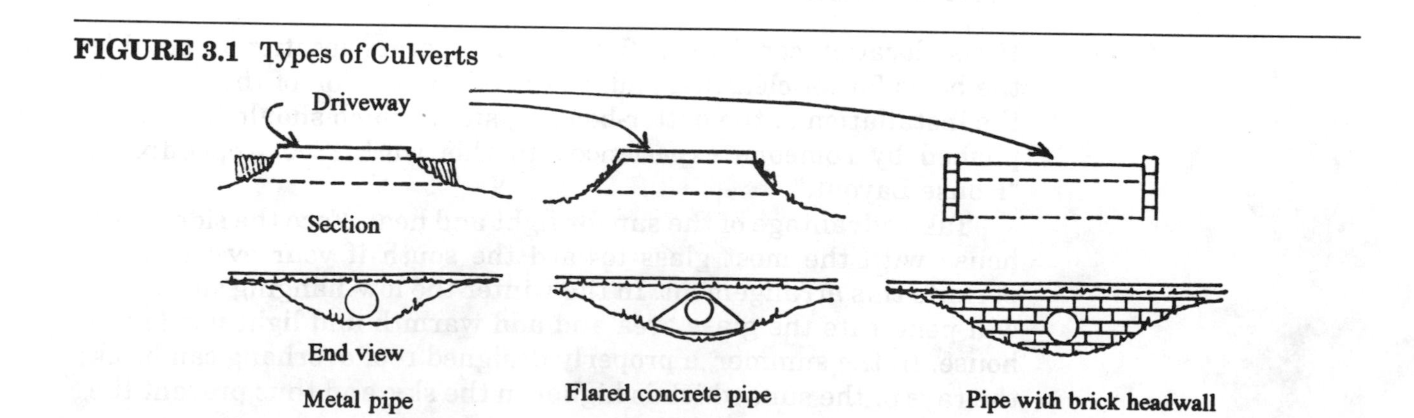

Topsoil

As part of the clearing, have the site contractor scrape off the topsoil in the construction area and stockpile it on the lot out of the way of construction. This topsoil can then be placed in the cleared area during the final grading.

Trees

All trees that are removed should be taken out completely, including the root system. You may be able to sell some of the trees to a local sawmill. Consider having the site work contractor cut them to length for use in a fireplace or stove, but be sure the cut logs are stacked where they will not interfere with the construction to follow, including the final grading. For more permanent storage later, place the cut logs on a frame of pipe or salt-treated wood several inches above the ground so that the fireplace logs will not be eaten by wood-boring insects such as termites.

Remove from the lot all stumps, tree limbs, brush, rocks and excess soil.

Flat Lot Problems

If your lot is very flat with little or no natural drainage, to avoid creating a moisture problem, bring in fill dirt to raise the lot elevation under the house above the final exterior grade (see Figure 3.2). This safeguard against standing-water problems after the

FIGURE 3.2 Foundation Buildup

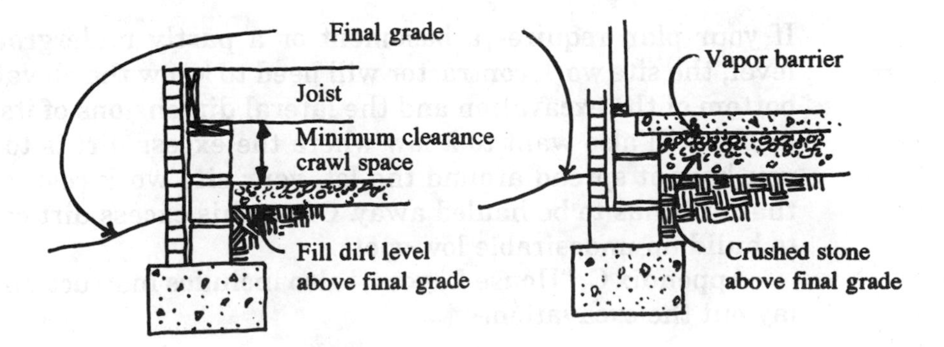

house is complete is especially important if your house is to be built with a crawl space. Check the local building codes to determine the minimum clearance required in the crawl space.

Starting the House

House Layout

The purpose of the layout of the house is to provide a measurement base on which the house is to be built by establishing the following:

- The exact lateral dimensions of the exterior walls
- The height of the floors and their levelness
- The squareness of the house to ensure that all corners are 90° or whatever angular dimension your plans call for.

The most widely used system to lay out a house involves the installation of a series of batter boards whose function is to hold in place a network of string lines.

The house layout task is somewhat complicated, and it should be carried out with skill. For readers who would like to tackle the job themselves, Appendix C, "House Layout," explains in detail an accurate procedure using simple tools that can be learned with the application of some study.

Should you decide not to do this work yourself, most surveyors and many lead carpenters have the skills to install a batter-board system.

Excavations

If your plan requires a basement or a partly underground lower level, the site work contractor will need to know the elevation of the bottom of the excavation and the lateral dimensions of its sides. He or she will also want to know where the excess dirt is to be put. If you have it spread around the lot, your site work cost will be less than if it has to be hauled away. Often this excess dirt can be used to build up undesirable low spots.

Appendix C, "House Layout," also includes instruction on how to lay out the excavation.

Sewage Disposal System

If your lot is served by a community sewage system, you only have to hook up to this system. This work should be included in the plumbing bid. It can be done at any time during the construction process as agreed upon between you and the plumber, but allow several weeks for settling of the ditch that carries the pipe from your house to the sewage system before final grading of the lot.

If no public or community sewage system is available, you will have to install on your lot an individual system designed for the local conditions, including the porosity of the soil. Among the possible systems are leaching cesspools, subsoil disposal drains and sand-filter systems. Whatever system you choose will probably have to be approved by the local health authorities.

Septic Tank System

One of the more widely used systems is the subsoil disposal method or "septic tank," as shown in Figure 3.3.

The principal components are:

- The concrete septic tank, which collects the sewage, decomposes the solids and passes the liquids on to the distribution box
- The small distribution box, which divides the liquid sewage and distributes it evenly among the drain lines

FIGURE 3.3 Septic Tank System

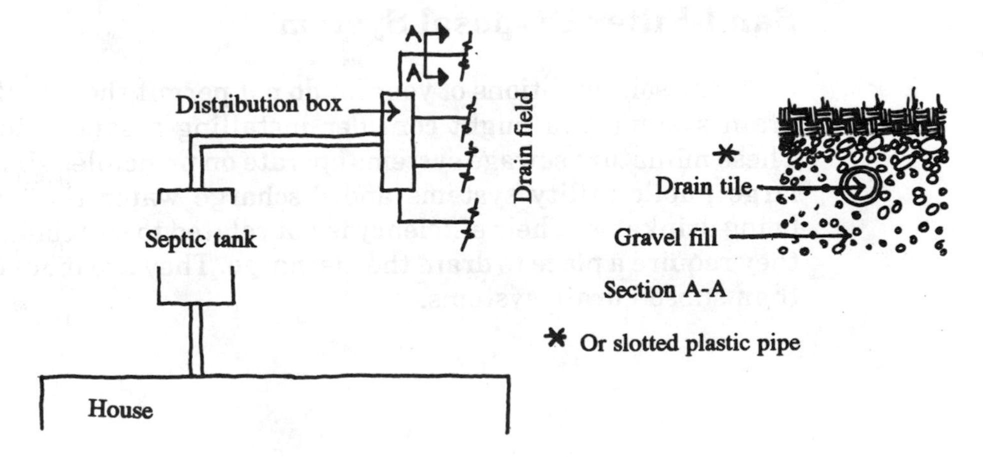

- The subsoil drain lines (drain field), which carry the liquid sewage to the various parts of the field where it is absorbed into the soil

The specifications for this system are usually determined by the local health department. After analyzing the soil and considering the size of the house (usually the number of bedrooms, which determines in large measure the number of people that could be expected to live in the house), the health department will issue a septic permit that specifies the size of the tank, the number of subsoil drain lines, their length and the width and depth of the trench holding each line.

The area to be used by the septic system should be cleared of all trees and other heavy growth to ensure that the system's capacity will not be reduced by future damage from roots. Weeping willow trees, in particular, have a way of working their roots into the pipelines and clogging them.

Check your lot size and house location to ensure that:

- The lot will provide enough room to accommodate the house and the septic system.
- If your plans require a well, there is enough room to provide a minimum lateral separation of 100′ between the well, if your plans include one, and the nearest point of the septic system, as required by most codes.
- The size of the lot will permit the movement of machinery (primarily a backhoe and a small dozer) to the septic area after the foundation of the house has been built. If this is not the case, install the septic system before you begin the foundation.

Sand-Filter Disposal System

If the soil conditions of your lot do not permit the use of a subsoil drain system, you might consider installing a sand-filter system. These miniature sewage systems operate on principles similar to the large public utility systems and discharge water that is close to being drinkable. Their efficiency is not related to soil conditions, but they require a place to drain the discharge. They are more expensive than subsoil drain systems.

Temporary Water Supply

Before masonry work can begin, you must have a source of water for mixing the mortar, cleaning the brick and other related masonry tasks. Check with your masonry contractor and determine what the requirements will be.

The following are various ways of supplying temporary water:

- Have the plumber install a temporary hookup to the local supply. This is a simple task if the supply line is already installed on or near your lot.
- Arrange to get water from a neighbor.
- Haul in water by truck using a 55-gallon drum or other suitable container.
- Install a well. Of course, this is to be done only if your plans call for a well as a permanent source of water. The masonry requirement for water, however, may dictate the timing of the well installation.

Temporary Electrical Hookup

Before the framing of the house can begin, you will need a temporary source of electricity for saws and other equipment. This hookup is normally installed by the electrical contractor, who will rig up a temporary service box on a pole, which must be inspected by the local building inspector. Then, the power company brings its lines to the box. This process may take two or three weeks, so planning for this service should begin early. In an emergency, a gasoline-powered generator will suffice, but in the long run it is better to have power on a more permanent basis.

Costing

- The cost of labor and materials for clearing, grading, excavating and laying the driveway base is included in the site contractor's bid.
- The cost of the culvert pipe is provided by the supplier and the labor for its installation is part of the site contractor's bid unless the highway department does the job. Labor for brick or stone head walls is contained in the mason's bid and the cost of the materials is provided by the masonry supplier.
- The cost of the materials for the house layout is provided by the lumber supplier and the labor is included in the bid of the

contractor given the job (or the owner if he or she decides to do it).

- The cost of the building permit and the water and sewer fees can be obtained from local building officials.
- The cost of the materials and labor for the septic or other disposal system is included in the bids of the trade contractors.

Management

- Get the building permit and pay fees for the sewer and water hookup.
- See that the appropriate trade contractors get the permits for electrical, plumbing and HVAC work. Post these permits on the site as required by local codes. Arrange for the site work contractor to begin the job.
- Line up other site work contractors (well digging and septic system) as needed.
- Minimize soil erosion by placing hay bales or other suitable material at probable water run-off areas. Disturb the original grade only as necessary for construction.
- Have the culvert pipe on site if the driveway needs one.
- Have house layout materials on site when needed.
- Start a daily log of construction events, including work done and materials delivered. It will be very useful in checking on progress and in keeping a record of materials ordered, including those not in the original estimate. It will be important for you to know when and if you exceed your budget and by how much.

4

Footings

The footing is the part of the structure that provides the base and contact with the ground for the house. Most footings are made of concrete reinforced with steel bar. In all-wood foundations, however, the footings are usually composed of gravel.

The two most important aspects of footing construction are that:

- They must be emplaced on firm, undisturbed soil. If your lot has been filled, the fill should be compacted by mechanical means.
- The top of the footing must be below the frost line. Check this requirement with the local building code.

Footing Size

If your house plans have been prepared for your lot by an architect or engineer, they will show in detail the size of the footings and the size and amount of reinforcing bar needed. If your plans came from a general source such as a plans service or paneled home manufacturer, the footings will not necessarily be suited for your lot and its soil conditions. It is very important that the footings and foundation be specifically designed for your lot. If this is not the case, get help from a professional architect or engineer. In many cases local building officials will advise you on code requirements for footings and foundations. As a general rule and as a *minimum* requirement:

- The footings for a one-story house should be not less than 16″ wide and 8″ deep.
- For a two-story house or a one-story house with a basement, the footings should be at least 24″ wide and 12″ deep.

These are minimums. Have a professional check your plan.

In all footings, whether or not the plans call for it, install reinforcing bar. This relatively inexpensive addition provides insurance against possible problems that can be quite expensive to correct after construction.

Figure 4.1 illustrates a typical footing consisting of poured concrete reinforced with steel bar. The sides of the excavated trench provide the forms for the concrete.

You will need to know the quantities of materials used in the footings since they are usually provided by the owner/general contractor.

Determine the amount of concrete by multiplying the depth × width × length of the footings and convert this figure to cubic yards. There are 27 cubic feet in a cubic yard. Do not forget to include the concrete pads under piers, fireplaces, columns and other areas shown on the blueprint. Having calculated the total amount of concrete, increase it by 25% to 30%. You will probably have about this much waste created primarily by digging out unexpected soft

FIGURE 4.1 Footings

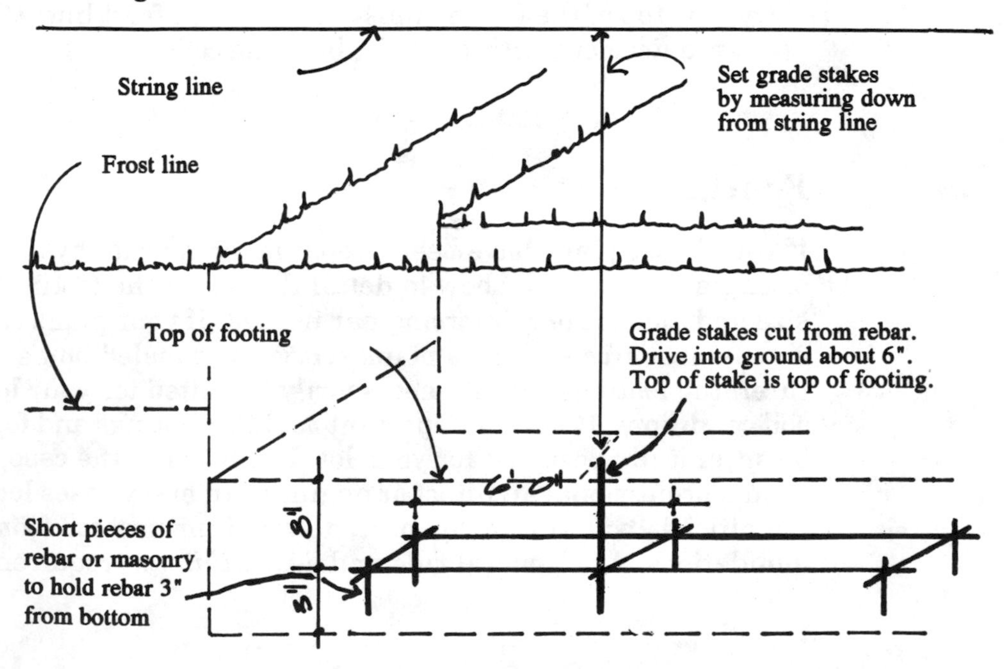

spots and some collapse of the trench side walls. Use 3,000 pounds per square inch (psi) concrete and order it by the cubic yard.

The reinforcing steel bar (rebar) requirement is based on the length of rebar needed. Allow at least a foot of overlap where the different bars meet. Rebar is usually supplied in 20′ lengths.

You will need stakes to hold the horizontal reinforcing rebar *within the bottom third of the concrete during the pouring.* Install these vertical supports at about 6′ intervals.

You will also need grade stakes to indicate the top of the concrete footing. Measure down from the batter-board string line the proper distance to establish the correct placement of the top of the stake.

Both grade stakes and the vertical supports for the horizontal rebar should be metal or masonry—for example, pieces of concrete block can be used to keep the rebar well up into the bottom third of the footing. This is not so good a system as the one using support stakes, because the rebar can slip off the piece of block during the pouring. Do not use wood. It will weaken the concrete, particularly after it has rotted away. Rebar is an excellent material for this purpose. Most suppliers will cut the rebar for you, so compute the number of grade and vertical support stakes needed before ordering the material and throw in a few extra. Vertical supports of about 8″ are all right, but the grade stakes should be long enough to allow about a foot to be driven into the ground plus the thickness of the footing. You will also need some malleable wire to tie the rebar to the vertical stakes.

Do not forget the rebar for the pads for the fireplace, piers, etc.

In addition, if your lot has any slope, you will need to install your footings with steps to compensate for the differences in grade (see Figure 4.2). For a block foundation, the height of each step should be 8″, the height of the block. To form these steps, you will need material such as plywood to hold the fresh concrete, as shown in the

FIGURE 4.2 Stepped Footings

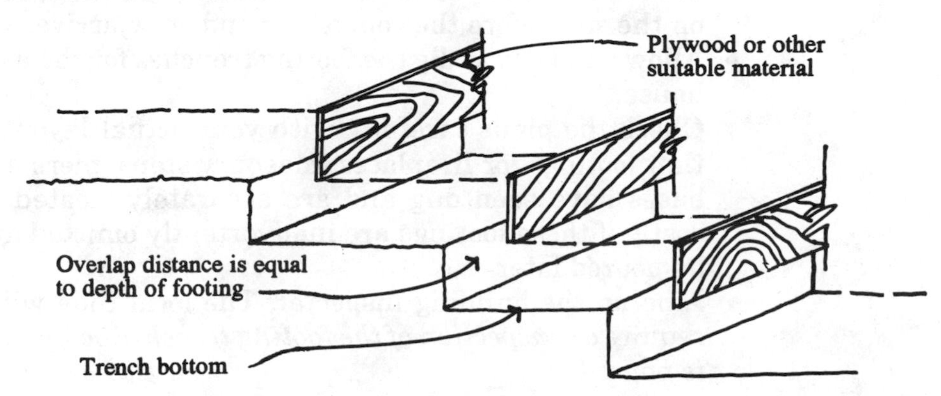

illustration. The plywood pieces should be 8″ high and about 4″ wider than the footing so that it can be forced into the dirt side walls of the footing trench for stability. You may need some small stakes (1″×2″×12″) to hold the plywood in place if the soil is soft. When pouring the concrete for these steps, it should be fairly stiff.

Footing Installation Plan

This job can be simple or complex depending on your house design and the lot. Whatever the case, you must have a plan for the footing operation. Consider the following:

- How will you get the concrete truck to the footing trenches? These trucks carry from seven to nine yards. Their flotation is not the best and they tend to be top-heavy. Get advice from your footing contractor. If your lot poses a very difficult problem in getting the concrete trucks to the footings, your concrete supplier may be able to offer additional advice.
- In some circumstances, the best way to get the concrete to certain areas of the footing trench is to move the truck across another trench already dug. If so, pack the inside of the trench with concrete block to prevent the side wall from caving in. When it is no longer necessary to cross this trench, remove the block and dress up the trench.
- On occasion the best means to reach the footing trenches is to build a makeshift wood trough, using it as an extension of the truck trough, or to use several wheelbarrows.
- In very troublesome situations, you may have to rent a concrete pumper at extra cost and pump the concrete into the footing trenches.

In all cases, have a workable plan before you order the concrete.

- Make sure that the material (less the concrete, of course) is on the site before the contractor and crew arrive.
- Allow a full day to dig the footing trenches for the average-size house.
- Check the plans carefully with your actual layout to ensure that footings for fireplaces, masonry steps, piers and column bases have been dug and are accurately located. It will be costly if these footings are inadvertently omitted and have to be poured later.
- Line up the building inspector. The local code will probably require *an inspection of the footing trenches before the concrete is poured*.

- Order the concrete when all inspections have been passed and the contractor is ready to pour.
- Once you have placed the order and the truck has left the yard, the concrete is yours. You usually have one to two hours from the time the truck leaves the plant until it must be emptied.
- Plan on pouring all of the concrete the same day unless you have an unusually large house. If the job requires more than one day, schedule the work so that concrete poured the second day does not link up structurally with the concrete poured the first day, which has already set up. For example, on the first day, pour the house footings; on the second day, pour the pads for the fireplace, piers, columns and masonry steps.
- Check on the weather to ensure that both precipitation and temperature forecasts are favorable. After the concrete has set four hours or more, rain should have little effect on the curing process. Concrete should not be poured when temperatures are expected to fall to near freezing. Don't forget that temperatures tend to drop after sundown.
- Allow at least two days for the concrete to gain sufficient strength before beginning the construction of the foundation.

Costing

Most contractors who specialize in footing work prefer to furnish only the labor and tools; the owner/general contractor supplies the concrete and other materials. If this is the case, be certain in your bidding that each contractor knows exactly what he or she is bidding on. You may want to ask for two bids, one for the complete job including the materials and the other for the job with you furnishing the materials.

Spell out your labor requirements in detail to the bidding contractors as follows:

1. Dig the footing trenches (and other areas such as fireplace base, column and pier pads). This can be done by hand or in part by machine (usually a backhoe) and finished by hand. Hand digging usually results in less waste of the concrete because there is less collapse of the walls. Machine digging may provide lower labor costs. The labor price should also include corrections to the excavations directed by the local building inspector in order to get approval.
2. Install the rebar, vertical supports and grade stakes, including tying this material.
3. Pour and distribute the concrete and finish it with a float (a wood trowel that levels the top with a rough finish).

Some footing contractors prefer to give a labor price based on a specified charge for each cubic yard of concrete poured. This method is acceptable, but it is preferable to get a fixed price for the job, because a price for each yard encourages the contractor to exaggerate the size of the footing trenches and thus increase both the labor and the material costs.

In summary, to get the total cost of the footings, both labor and material costs can be provided by the contractor, or the labor cost can be provided by the contractor with the material furnished by you from the supplier. You will need to know the quantities of concrete by the cubic yard, rebar to include the grade stakes and vertical supports, tie wire for the rebar and the material for the steps, if any.

Management

- Make sure that the footing trenches are dug along the correct lines. Although this sounds ridiculous, mistakes do happen. To avoid this problem, mark the ground under the string lines where the footing trenches are to be dug. Use colored plastic tape or lime.
- Make sure that all loose spoil is removed from the trenches.
- Check to see that the bottoms of the trenches are down to firm, undisturbed soil. Be particularly wary of areas that originally contained a large tree root system. If the soil is still soft, continue digging until firm soil is reached.
- Make sure that the sides of the trench are more or less vertical.
- Be sure that the foreman knows where you want the spoil thrown. Without being told otherwise, most crews will place the spoil where the least effort is required. Generally, spoil should be thrown outside the house area, where it can be easily handled by machinery when the final grading is done. If spoil is placed inside the house area in which a concrete slab is to be the first floor, it may interfere with the preparations needed to give the slab a good base. If the plans call for a crawl space, piles of dirt will interfere with the work of crews who will be moving around in the crawl space later.
- Before the concrete is poured, make one final check of the trench alignment by placing the batter board string lines in place to mark the foundation location. Then check the alignment of the trenches.

Be aware of the tendency for workers to add too much water to the concrete so that it will flow easily through the trenches and thus require less labor to place. Most concrete trucks are equipped with water tanks to permit thinning. Although some modest thinning may be required, thinning to the extent the concrete flows easily will drastically reduce its final strength. If required, water should be added only to the extent that the concrete mix is still able to stand without spreading out flat due to its own weight. There should be no visible pools of water in the concrete. With the proper consistency, most of the concrete will have to be more or less placed in the trenches by constant moving of the truck trough, by shoveling, by wheelbarrow or by a combination of all three.

A more accurate method of properly controlling the concrete-water mixture is the slump test. This test is performed on the concrete just before it is poured at the site. It calls for a metal cone open at both ends with a base diameter of 8″, a top diameter of 4″ and a height of 12″. The cone, provided by the concrete supplier, is filled with concrete from the truck and packed tightly to remove all pockets. The cone is then removed and placed alongside the concrete. Without the cone the concrete will slump or fall to some degree. The amount of the slump is measured: if it falls between 2″ and 4″ go ahead and pour. If it falls more than this amount, the concrete has too much water in it. If the fall is less than 2″ it is going to be difficult to work and needs a little more water.

If loose dirt should fall into the concrete, have it removed with a shovel. The dirt will weaken the concrete at that point.

Your computations for the amount of concrete needed may not be the same as the actual requirement. For example, running into large amounts of soft soil will call for more excavation and thus more concrete than planned. Keep the concrete moving to the job, but hold off ordering the last truck load until all other trucks have been emptied. Then estimate how much more you will need and order that amount. You should be able to keep the waste to a half a yard or less.

Keep the concrete supplier's trip tickets. They will give you the exact number of cubic yards of concrete delivered and provide a firm record for labor payment if your contract is on the per-cubic-yard basis.

5

Foundations

If your plans were prepared specifically for your lot by an engineer or an architect, they should show all the necessary details of construction, including required reinforcing.

On the other hand, if your plans were provided by a plans service, it is unlikely that the foundation design fits your lot. If the plans and shape of the lot permit foundation design not exceeding four or five courses of block, you should not need any extra strengthening. Should your situation require a full or partial basement (particularly if this is a modification of the original plan) or if the lot slopes steeply and your crawl space requires walls of six or more courses of block, it is probable that additional reinforcing is needed. You should consult with a professional architect or engineer to get safe design criteria for your foundation.

The following are samples of the more commonly used reinforcement systems for higher foundation walls:

FIGURE 5.1 Horizontal Block Reinforcing

- *Horizontal reinforcing* (see Figure 5.1) is applied as the mason lays the foundation. It can be used for each course (it is buried in the mortar), every other course or just every third course. Generally, the greater the backfill that will be placed against the wall, the more horizontal reinforcement is required. Horizontal reinforcing can be supplied in different widths to match the widths of the block or block-brick combination used in the foundation wall.
- *Vertical reinforcing* (see Figure 5.2) is applied in two commonly used methods. One is the use of reinforcing bar and concrete to build a solid steel and concrete column. The other is to build into the walls pilasters of extra block to give the wall greater thickness at selected points.

If your plans are based on a poured concrete foundation, the above remarks apply equally as well. In this case, however, the horizontal reinforcing will most likely be made of rebar rather than the special truss designed for block.

Check with your local contractors to determine which type of foundation, concrete or block, is best for you. Block is usually better if your plans call for a lot of corners in the foundation. It is also the superior material if your plans require many openings in the foundation for doors and windows. On the other hand, concrete is usually stronger and more waterproof (it will need extra waterproofing, as discussed later in this chapter). Concrete becomes more important if you have underground living space in your plan, but may be more costly.

FIGURE 5.2 Vertical Block Reinforcing

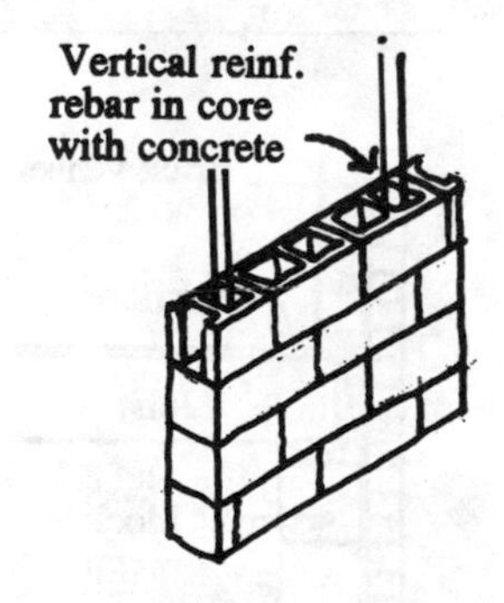

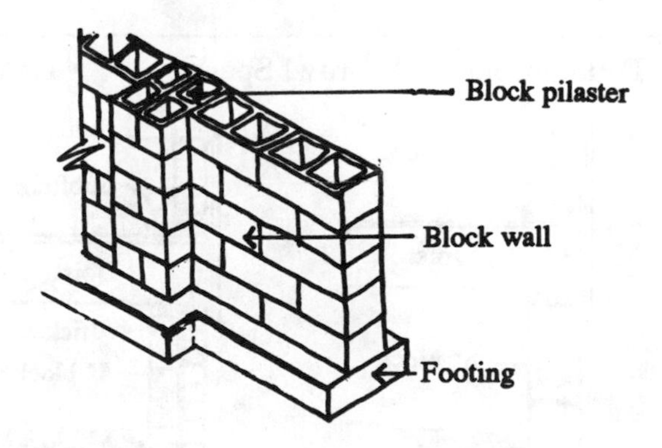

The Brick/Block Crawl-Space Foundation

The crawl-space foundation made of block or a combination of brick and block has three forms:

1. **The first method is the use of block by itself. The exterior of the block can be left as is, or it can be painted or stuccoed.**
2. **The second form is the block/brick combination, where the brick forms part of the load-bearing foundation and thus supports the wood-framing structure in conjunction with the block. This form of foundation is best used when the exterior walls are to be finished with wood, aluminum or vinyl siding.**
3. **The third form is used when the exterior wall is to be completely covered with the brick veneer as a finished surface. The brick of the exterior walls are actually an extension of the brick veneer of the foundation wall. The brick does not support the structure. Actually, the structure (foundation wall and wood frame) supports the brick.**

Crawl spaces must be vented to allow the circulation of air to the outside. Foundation vents, usually made of aluminum, are 8″×16″, the size of the concrete block. During cold weather these vents can be closed to retain the heat in the crawl space.

Figure 5.3 shows three examples of foundations with crawl spaces. The number of vents and their spacing is usually indicated in the blueprint. As a general rule, at least one vent should be close to each corner of the house and there should be cross-ventilation for

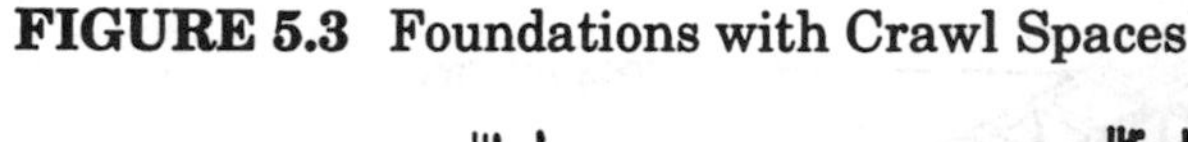

FIGURE 5.3 Foundations with Crawl Spaces

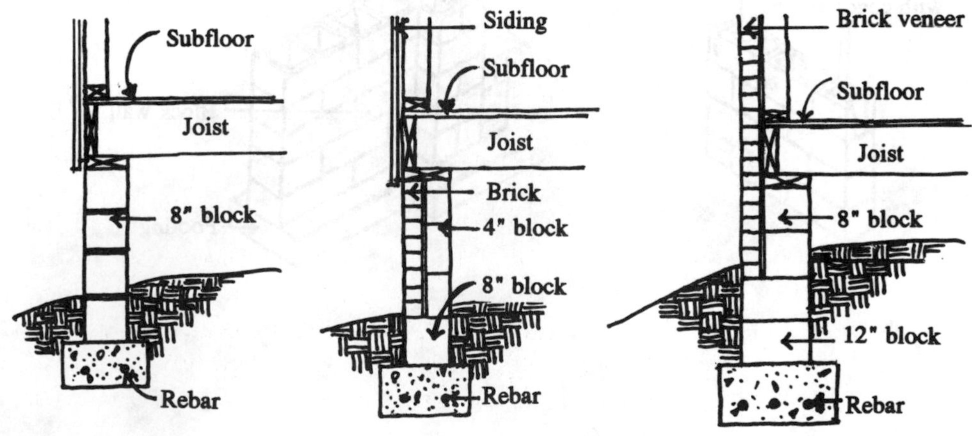

at least two opposite sides. If you live in an area of high humidity, install at least one vent for each 15 linear feet of exterior foundation wall.

In addition, access to crawl spaces must be provided by doors not less than 18″×24″. Ready-made steel doors of various sizes can usually be found at the masonry suppliers. These doors can also be made of wood by the carpentry crew during the framing job. Use salt-treated wood.

If your crawl space is large (over 2,000 square feet) or if the shape is unusual, you should consider installing more than one crawl space access door for the convenience of the crews who must work in this area during construction and later when repair or servicing of the house requires work in the crawl space.

The Basement Foundation

Figure 5.4 details the various elements of the basement foundation. The basement foundation wall can be built in the same three forms shown in Figure 5.3.

Waterproofing material should be applied over the exterior of the basement wall below grade and a slotted drain pipe in a gravel bed added at the base of the footings to remove the water and prevent pressure buildup.

Waterproofing can consist of several layers of tar and building paper or several coats of waterproof masonry material or a

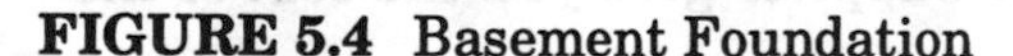

FIGURE 5.4 Basement Foundation

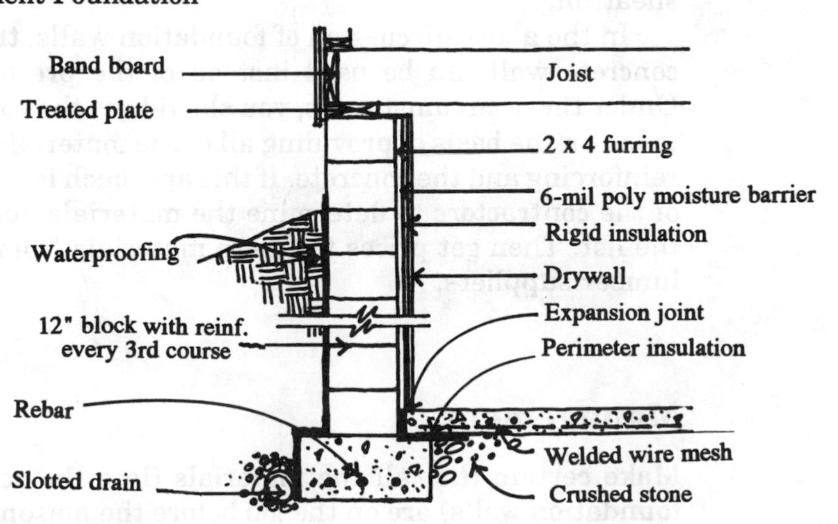

combination of both. A recently developed, more effective waterproofing material is a synthetic rubber waterproofing membrane. For use over concrete block, it is best to first apply parging over the block, then follow with the synthetic rubber sheeting after the parging has dried.

The waterproofing must be protected during the backfilling process or rocks and other hard material in the backfill can penetrate it and allow water seepage into the living area. Large sheets (4′×8′ or 10′) of impregnated sheathing (an inexpensive compressed fiber material ½″ thick) are excellent for this purpose and should be applied against the waterproofing before the backfilling is begun.

Backfilling must not be done until the floor framing of the first floor has been completed, including the laying of the subfloor. Without this bracing, the foundation wall could collapse.

Costing

The cost of labor for the foundation walls is provided by the mason, who may also prefer to furnish the materials as well. If not, get the supplier to "take off" (determine the quantities needed) the foundation materials and price them.

If your choice of waterproofing is the building paper/tar system or the rubber membrane, most roofers will give you the cost of both labor and materials. If one or more coatings of cement based waterproofing compound is required instead, the labor can be priced by the mason and the material by the masonry supplier.

The backfilling is usually performed by the site work contractor with the owner–general contractor furnishing the impregnated sheathing. Have the lumber supplier give you the cost of the sheathing.

In the above discussion of foundation walls, the poured-in-place concrete wall can be used instead of the precast concrete block. Under these circumstances, you should get the concrete contractors to bid on the basis of providing all of the materials for the forms, the reinforcing and the concrete. If this approach is not feasible, ask one of the contractors to determine the materials needed and give you the list. Then get prices for these materials from the masonry and lumber suppliers.

Management

Make certain that all the materials (less the concrete for poured foundation walls) are on the job before the masonry crew arrives. If your contract calls for the mason to furnish the materials, it is his or her responsibility to have them on the job.

Visit the job at least once a day. Survey the material and talk to the head mason to be sure that the supplies are adequate. For some strange reason, many contractors realize they are running out of materials only when they actually do so. This usually means the loss of one or more days' work.

Before the mason begins laying the block or brick, be sure that the batter-board string lines are installed in the proper slot. The footings contractor has probably cut additional slots to measure the outside limits of the footings. This can cause the mason to begin laying the block on the wrong line position.

The following is a list of materials needed:

Block	Horizontal reinforcing
Brick	Vertical reinforcing
Mortar cement	Anchor bolts
Masonry sand	Concrete and steel lintels
Brick ties	Crawl-space door
Foundation vents	Window, door frames for basement

Anchor bolts are set into the masonry with enough of the threaded end projecting upward through the wood sill plate so that a washer and bolt can be installed to hold the sill plate to the masonry wall. Lintels are devices used to span openings in masonry work for doors and windows (see Figure 5.5 and Figure 5.6).

If your foundation has these openings, you will need a concrete lintel to span each one. The simplest type of concrete lintel is the precast version, which you can get from the masonry supplier. Its

FIGURE 5.5 Masonry Lintels

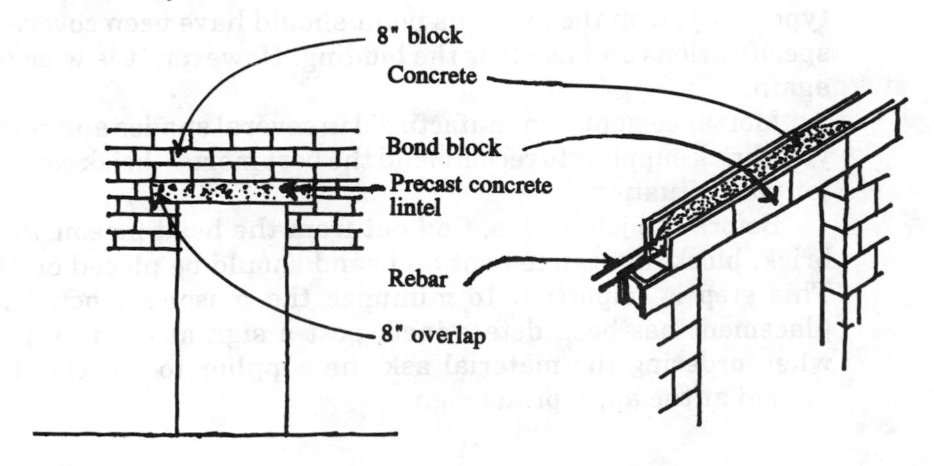

FIGURE 5.6 Steel Lintels

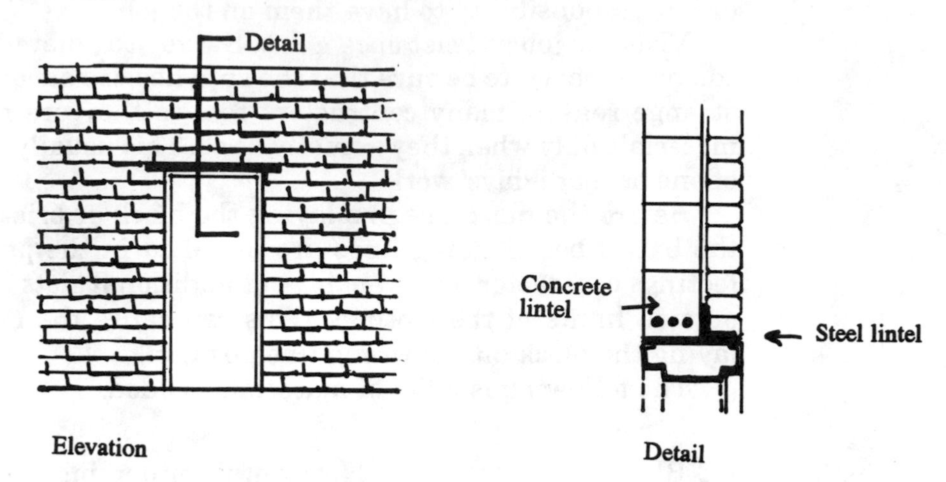

height and width are matched to the block and its length should allow at least an 8″ overlap on the bearing blocks on each side.

Another method is illustrated in which the lintel is made on the job using a special bond block formed like a "U". It is much more complicated than the precast lintel, requiring rebar and a wood form to hold the bond blocks in place until the concrete has set.

If your wall has a brick veneer over the block, you will need steel lintels to support the brick over the door or window opening.

Steel angle iron with sides that are wide enough to support the brick is fine for this job. With standard brick, an angle iron with 3½″ legs works well. Again, allow at least an 8″ overlap on the supporting brick below.

Discuss with the head mason the type of mortar joint you want in the brick and block. Some of the types available are flush, struck, concave, beaded, weathered and grapevine (for colonial work—it requires a special tool). The head mason can illustrate the various types for you on the job. This point should have been covered in the specifications and used for the bidding. However, it is wise to check again.

Mortar cement is manufactured in several shades and colors. Ask your brick supplier to recommend the best mortar-brick combination for your situation.

Before the job begins, find out from the head mason where the brick, block, mortar cement and sand should be placed on the site. This step is important to minimize the mason's labor. Once the placement has been determined, post a sign at each location and when ordering the material ask the supplier to tell the driver to unload at the appropriate sign.

Brick and block should not be laid when they are wet or when temperatures are near freezing. Provide some sort of cover at the job site so that this material can be protected when not being used. Polyethylene sheet material is excellent for this purpose.

Although mortar cement is packaged in what is supposed to be waterproof bags, the cement should have additional protection. Most suppliers deliver mortar on a standard wood pallet. Ask them to include a waterproof cover with each pallet. It is worth the extra cost.

The trucks that deliver the brick and block are very heavy and are equipped with special unloading cranes. It is important that they be given sufficient space for operation. Discuss this problem with your mason, who should be familiar with these trucks and their requirements. If your lot has little space for storing materials, consider having the material delivered in more than one trip with a day or two in between to allow the mason to use up the supplies on the job and release storage space for the additional material.

If your plans require windows and doors in the foundation walls of a basement, have them on hand so that they can be installed by the mason. Having them on site during the masonry work also permits the mason to check the openings required in the masonry with the items to be installed.

The Block/Slab-Floor Foundation

Figure 5.7 illustrates the block foundation wall with a concrete-slab floor poured over compacted fill.

In addition to the obvious differences with the crawl-space design, note the following:

FIGURE 5.7 Block Foundation with Slab-Floor on Grade

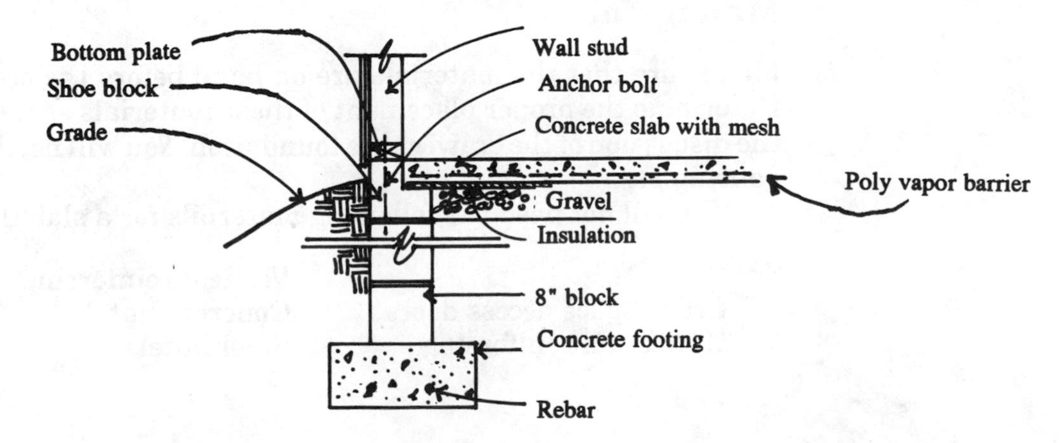

- The top course of the block is a header or shoe block in which a portion of the block has been cut out by the manufacturer to provide a base for the concrete slab.
- The slab is laid on the top of the rigid perimeter insulation (see Chapter 12, "Insulation").
- There is a vapor barrier of 6-mil polyethylene below the concrete and on top of the fill.

Compared with the crawl-space design, the slab requires less insulation, and it is more solid and less prone to problems from ground moisture. On the other hand, the crawl-space design will make later changes to the structure much easier; it is readily adaptable to a nonlevel lot and is less tiring on the feet.

Although the illustration shows the foundation wall as concrete block, this same design can be built with a combination brick/block foundation wall. Use the brick as the form for holding the slab floor rather than the header block. It can also be modified for a complete exterior brick wall, as shown in Figure 5.3. The header block is used, but a base for the brick veneer is needed, either by laying the brick on the footing or by laying a larger block for the first course or two and then laying the brick on top of this wider block.

Costing

The cost of the labor and materials for the block and brick is provided as indicated in the discussion of the crawl-space foundation.

The cost for labor and materials for the concrete work is given in the bids of the concrete contractors. Again, you may find that the contractors prefer that the owner–general contractor furnish the materials. If so, have the masonry supplier give you a take-off of the materials and provide a price for furnishing them.

Management

Make sure that the materials are on hand before the crew arrives. Determine the proper placement of these materials as mentioned in the discussion of the crawl space foundation. You will need a concrete pouring plan.

You will *not* need the following materials for a slab floor:

Vents
Crawl space access doors
Horizontal reinforcing
Vertical reinforcing
Concrete lintels
Steel lintels

Additional material needed for the slab:

Wire mesh (6 gauge, 6×6)
Fill material
Polyethylene
Concrete
Rigid insulation

The order of the construction after the footings have been poured is as follows:

1. The foundation is laid.
2. The under-slab fill is brought in and compacted.
3. Any rough-in of plumbing, electrical and HVAC is installed *and inspected by the local building inspector.*
4. The poly vapor barrier is put down.
5. The perimeter insulation is installed.
6. Wire mesh is laid.
7. *The work is inspected again.*
8. The concrete is poured and finished.

If the amount of fill is substantial (more than a 4″ layer of sand or gravel), have the site contractor build up most of the grade before the footings and foundation walls have been built. The concrete crew will spread and compact the final top layer of the fill, including any spoil from the under-slab mechanical work, dress it up and lay the floor.

If the concrete contractor uses wood-grade stakes, which are needed to provide level points throughout the slab area, make sure that they are removed and the holes filled with concrete during the final finishing. Wood will eventually rot and weaken the slab.

The footings receive only a float finish, but the slab must be given a trowel finish in most cases, which is much smoother and requires the concrete to be more "set up"; consequently it will take more time. Consider this point in your scheduling. Start early in the morning.

The slab finish is much more sensitive to damage from rain or freezing precipitation than the footings. Watch the weather forecast. Have additional polyethylene on hand to cover the slab after it has been finished. This covering will not only protect the slab from the weather but it will also provide better curing since it keeps the moisture in the concrete longer.

The Integrated Concrete Footing and Slab

Figure 5.8 illustrates a combination of foundation and footings in a single piece of concrete. This method is simple and relatively inexpensive. The design should only be used on a flat lot. Even then it

FIGURE 5.8 Integrated Slab Foundation/Footing

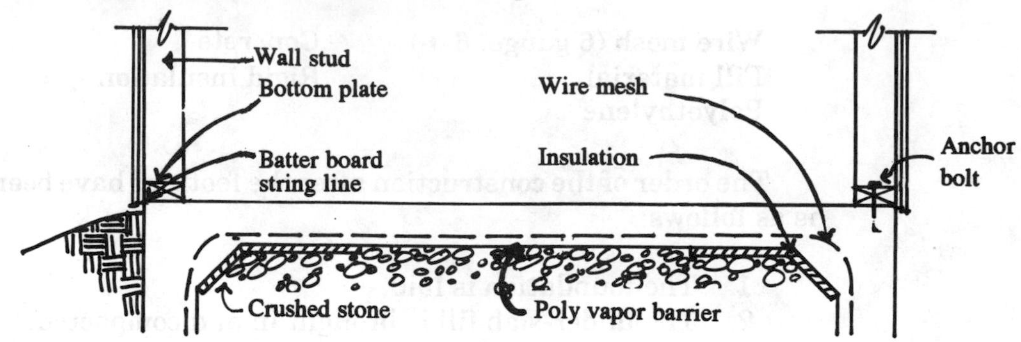

usually requires some site work to prepare the ground for receiving the concrete.

Once the site work has been completed to provide the proper grade and compaction, the batter boards are installed. The next step is to dig the footing trenches. Since the footings and foundation walls are one piece of concrete, the string lines of the batter boards show the exterior line of both the footing and the foundation. The lines indicate the level and dimensions of the finished floor.

The next step is to install the forms for the outside of the slab that will appear above grade. At this stage any under-slab plumbing, heating and electrical work is installed by the trade contractors. The grade for the slab is dressed up in preparation for the concrete pour. The rebar, mesh, insulation and vapor barrier are then put down, *all work is checked by the building inspector* and, finally, the concrete is poured.

Costing

The cost of preparation of the grade, including both labor and materials, should be part of the site work contractor's bidding. It is much less costly to do this type of work by machine and the final dressing by hand.

Again, in determining the costs for the concrete work, you have two choices as to who provides the material. In this case, however, it is to your advantage to have the concrete contractor provide the form material. If you furnish this material, it will be used only once and discarded. The concrete contractor, who uses the forms over and over again, has a much lower cost for these items.

Management

- Have on hand the following materials:

Fill material	Form material
Rebar and tie wire	Polyethylene
Wire mesh (6 gauge, 6×6)	Anchor bolts
Rigid insulation	

- *When ordering concrete for slabs, specify that fiber mesh material be added. It will effectively reduce cracking in the finished product.*
- Have a plan for getting the concrete trucks to the job.
- Make certain that the under-slab plumbing, heating and electrical work has been completed and that the footings and slab work are checked by the inspector before the concrete is poured.
- Keep track of the weather and avoid pouring on days of expected freezing or near-freezing temperatures and expected precipitation. A common rule of thumb is to pour if the temperature is 40° F and rising.
- Plan to start the pour early in the morning. The work should be continuous until the trowel finish is complete.
- Cure the floor after the troweling by covering completely with polyethylene or coat with a spray-on curing compound. Leave this coat on for at least three days. In very dry climates, keep the concrete moist by spraying with water as needed.

The All-Wood Foundation

Figure 5.9 illustrates both the crawl-space and full-basement versions of the all-wood foundation.

This method of building footings and foundations uses gravel and treated wood. It usually requires less time than the more conventional masonry method and should cost less. It can be installed in temperatures below freezing and thus permits construction in weather conditions that would halt concrete and masonry work.

The exact, proper construction of this type of foundation and footing is very important, particularly if the plan calls for a below-grade living area or basement.

The wood used in the construction must be salt-pressure treated and the below-grade fasteners (nails and bolts) must be stainless steel. The pressure treatment for all lumber below grade should be .60 CCA (pressure treated lumber with .60 pounds of Chromate Copper Arsenate salt solution per cubic foot of lumber. See page 63),

FIGURE 5.9 All-Wood Foundation

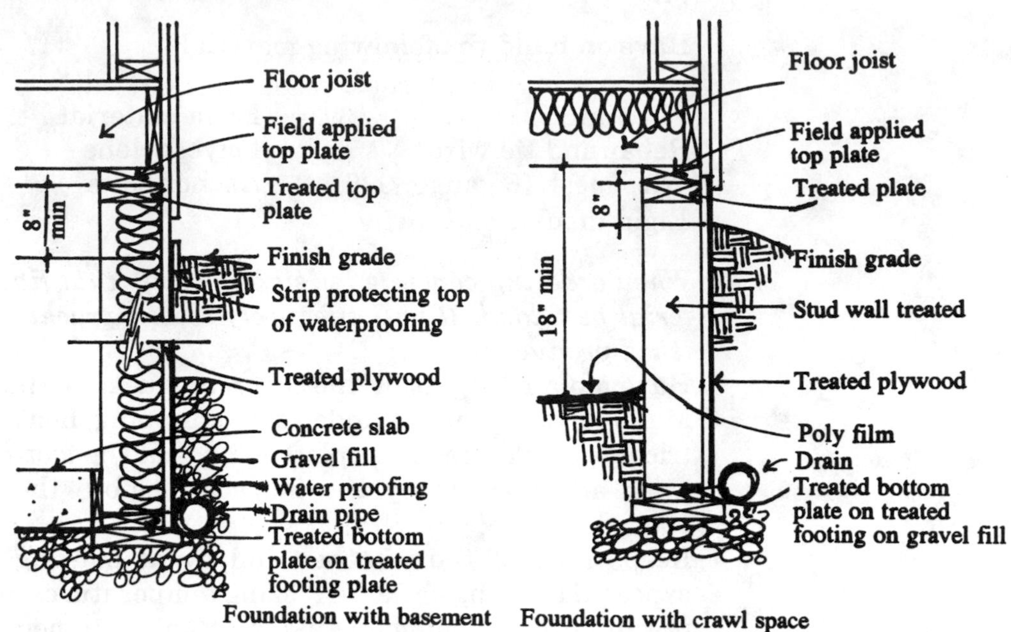

which is 50% higher than that recommended for normal ground contact.

The key element in this type of foundation is the control of moisture. Roof water runoff must be directed away from the foundation by properly installed gutters, downspouts and splash blocks or, preferably, an underground pipe drain system. The joints between the plywood panels must be caulked. Final grading of the lot must be sloped away from the foundation, as it should be in any foundation construction.

In the basement version, plywood joints must be caulked full length and covered with a 6-mil polyethylene film to direct the water down into the footing drain system. As an added precaution, coat the foundation over the polyethylene with two coats of hot or cold tar with alternating layers of building paper or, preferably, coat the plywood with the rubber waterproofing membrane previously mentioned.

Footings can be gravel, crushed stone or sand at a minimum thickness of 4″ for the crawl space walls of a one-story house. For basement walls and two-story houses on crawl spaces, increase the thickness to 6″. Note in the basement version the need for the extra gravel around the base of the foundation and the drain-pipe system.

Costing

The labor cost for the footings is provided by the footing contractor with the owner–general contractor providing the gravel. In the basement version, the costs of labor and material should include the larger amounts of gravel and the drain-pipe system (corrugated plastic slotted pipe is fine) to be installed after the wood foundation walls are up and waterproofed.

The cost of the labor to install the wood foundation should be included in the bid of the framing contractor with the lumber supplier doing the take-off for quantities and price.

Management

- The process of installing the footings is similar to that for concrete except that there is no requirement for rebar and the grade stakes can be treated wood.
- Determine with your footing contractor how to handle the gravel. Should the entire load be dumped at one place, in several places or from the truck directly into the footing trenches?
- Check with the local building officials with regard to inspections needed for the footings and any special requirements.

For management tips about the carpentry work, see Chapter 6, "Framing."

The Bonded-Block-Foundation Wall

Bonded block is a method of building a block wall without using conventional mortar (see Figure 5.10). If you choose this system, construct the concrete footings in the usual way, then have your mason lay only the first course of block using mortar in the normal manner. The rest of the foundation wall is then laid dry without the use of any mortar. To maintain a plumb, level wall, small amounts of sand can be inserted to raise blocks where required to keep the alignment.

When the dry laying of the block is complete, *both* vertical sides of the wall are covered with a thin layer (about ⅜″) of a special mortar consisting of hydrated lime, Portland cement and a special fiberglass material. Most masonry suppliers carry this material. Follow the instructions of the manufacturer and you should have no trouble.

FIGURE 5.10 Block-Bond Wall

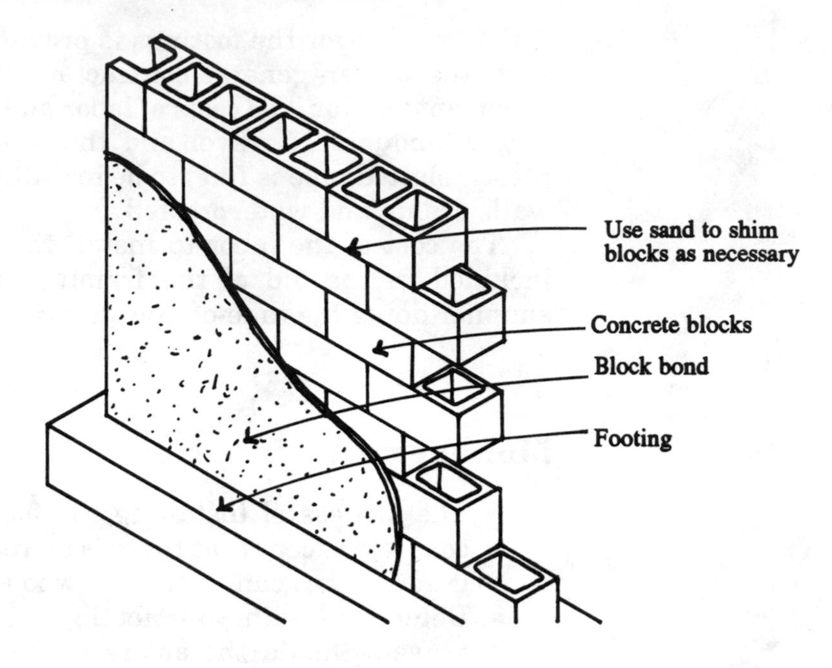

The advantages of this system are that it requires less skill and the wall is made waterproof by the application of the masonry and fiberglass coating. The strength of the wall is at least equal to that of one laid fully with mortar. Horizontal reinforcing is not normally required and the wall will present an attractive stucco-like finish that is easily painted. This wall should not be covered with brick veneer. Since the mortar is missing from the joints of the block, the top of the brick veneer and the block will almost always be of different levels.

Costing

The cost of the block and the fiberglass binding material is provided by the masonry supplier with the labor cost furnished by the masonry contractor. Since little mortar is used the block requirements will increase. The information sheet published by the bonding material manufacturer gives instructions on how to compute the additional block and the amount of bonding material needed.

Management

Management tips discussed in the previous sections will apply here as well.

Termite Control

Termites can cause serious damage to houses if proper precautions are not taken. The most effective means of controlling termites is the poisoning of the soil around the foundation of the house by a professional contractor and the use of salt-treated wood in areas where the wood could be exposed to moisture by contact with the ground, damp concrete or brick or block. See Chapter 6, "Framing."

Have the contractor include a guarantee for five years, which in essence says that the contractor will pay for any damage to the house caused by termites and other wood-boring insects that occurs in that period. At the end of the five years, this guarantee can be extended for an additional fee.

Costing

The complete cost for termite control, including labor, materials and the guarantee, is included in the contractor's bid.

Management

If your house plan is based on crawl-space construction, notify the contractor to apply the soil poisoning after the foundation is complete and the crawl-space grade is finished.

If your house plan is based on concrete-slab construction, notify the contractor to apply the soil poisoning after the foundation is complete, all under-the-slab plumbing, electrical and heating work has been installed and the grade for the slab has been finished but before the concrete slab has been poured.

6

Framing

The wood frame forms the shape and size and provides the strength of the structure. In a brick house (in modern construction this is actually brick veneer), the framing supports the brick, not the reverse, as many people think. It is very important that the framing be erected correctly. It is not an area in which to compromise to reduce costs. Errors discovered after the framing is complete can be expensive to correct.

Although it is not the purpose of this book to teach you how to frame a house, you should familiarize yourself with the framing system of your plans so that you know the major features and the names of the various parts. This knowledge will be very useful in talking to your suppliers and to your carpentry crew.

Lumber Selection

There are generally three choices in the selection of lumber for framing: fir, SPF (spruce, pine and fir mixture) and southern yellow pine (SYP). Fir is an excellent lumber. It is strong and easy to work, but it can be expensive and not readily available in your area.

SPF is a mixture of several species of lumber and generally consists of various proportions of white spruce, Engleman spruce, lodge pole pine, and Alpine fir. Expect the bulk of the wood to be spruce. SPF is a moderately strong lumber, but less so than either fir or SYP and may require, in certain load-bearing roles, a larger structural dimension with increased costs. SPF lumber works very well, so you may save on labor costs. Like fir, it is very stable and will not usually twist or warp after installation. SPF should be available in most areas of the country.

SYP is a strong species that is less costly than the two options described above. It is harder to work, however, and may cause labor costs to be higher. SYP also has a greater tendency to twist and warp after installation in the framing system. With proper blocking (a possible additional expense), it will usually perform within acceptable limits.

Considering the pros and cons of these types of lumber, a good compromise is a selection of fir or SPF for studs and top and bottom plates with SYP for the joists, rafters, headers and all other elements of the system. All lumber should be kiln dried (KD).

For those parts of the framing in which the lumber is to be exposed to moisture, and damage from termites and other wood-boring insects, you should use pressure-treated lumber to improve its life. There are three different degrees of treatment as follows:

- .25 pounds of salt solution per cubic foot of lumber for above grade such as decks and railings
- .40 pounds of salt solution per cubic foot for lumber in contact with the ground, such as columns supporting decks
- .60 pounds of salt solution per cubic foot for lumber below grade, such as the all-wood foundation

Wet salt-treated wood should be handled with gloves and a respirator or dust mask should be worn when using a power saw to cut it. *Scrap pieces should not be burned. Their fumes are toxic.*

Framing Designs

- *Platform framing* techniques are shown in Figure 6.1. In this concept, each floor is built separately with the first floor providing a work platform for the structure to be built above. Platform framing is the most widely used system.
- *Balloon framing* is illustrated in Figure 6.2. Note that the studs of the exterior walls are continuous from the sill plate of the first floor to the top plates of the second floor. This type of framing is more costly in material since one 18′ stud is more expensive than two standard studs of 93″. Labor costs are also higher, since the structure requires much greater use of scaffolding.
- *Plank-and-beam framing* is shown in Figure 6.3. Note that this type of framing uses greater spans for lumber and requires fewer pieces, but with greater dimensions. It is very

FIGURE 6.1 Platform Framing

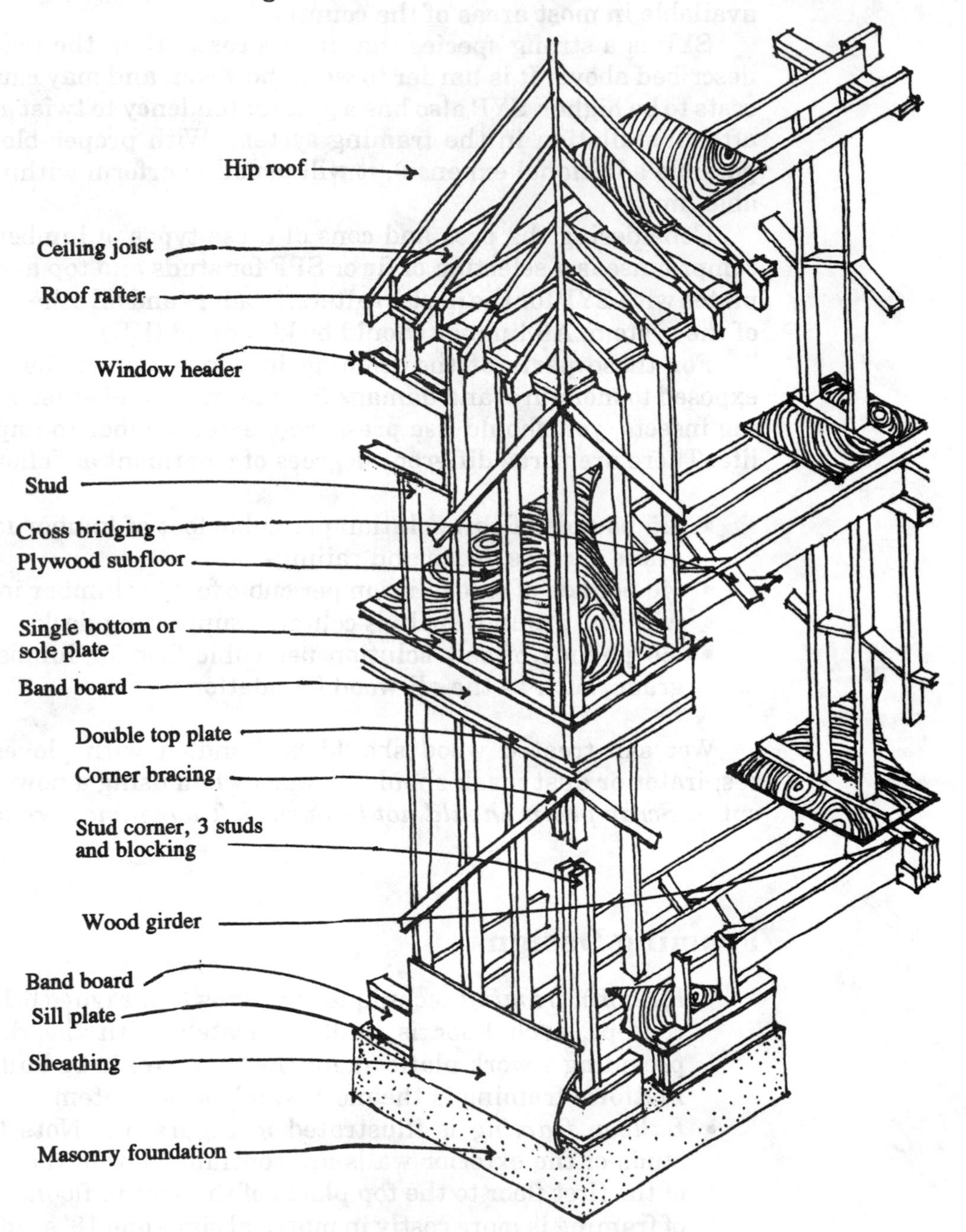

FIGURE 6.2 Balloon Framing

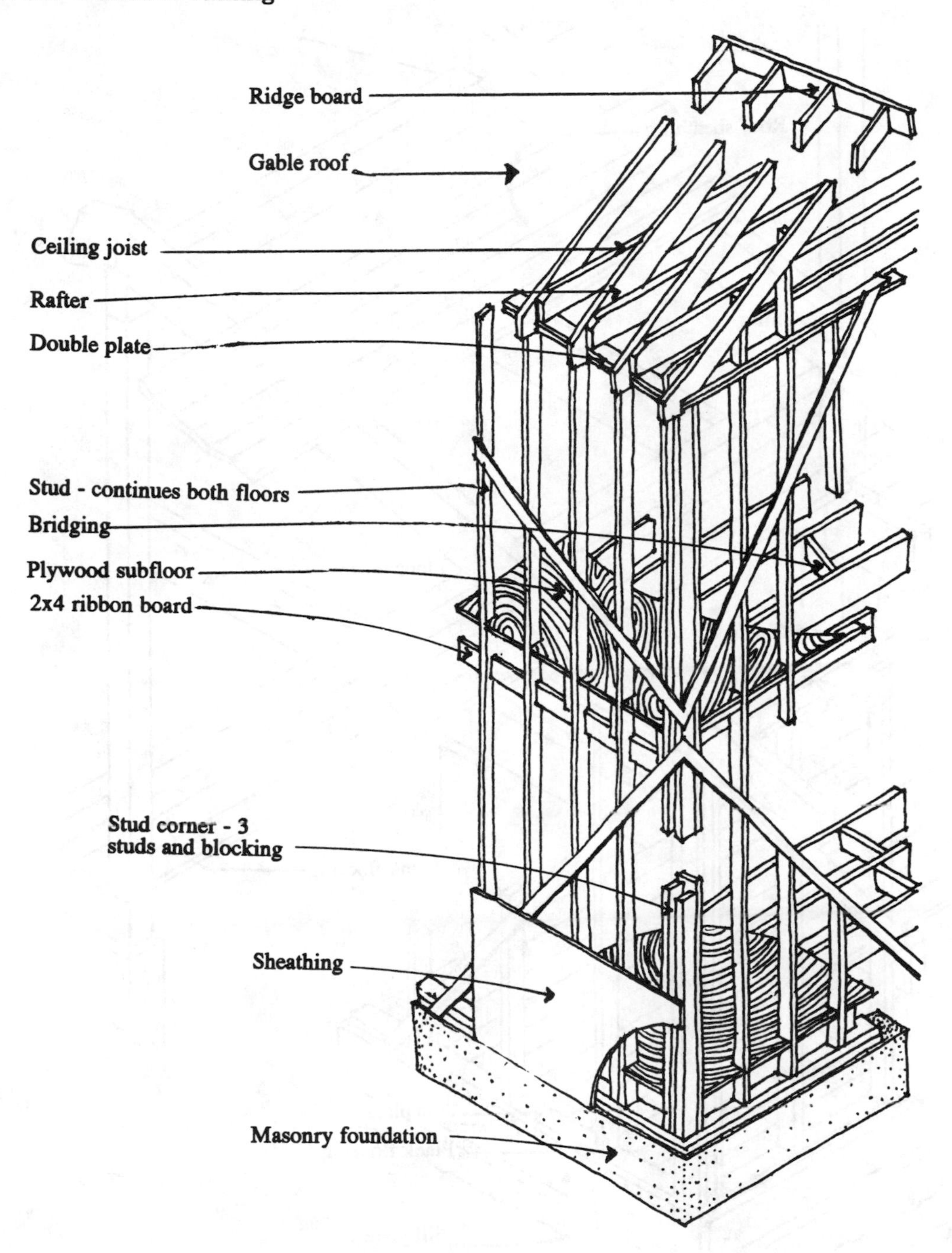

FIGURE 6.3 Plank-and-Beam Framing

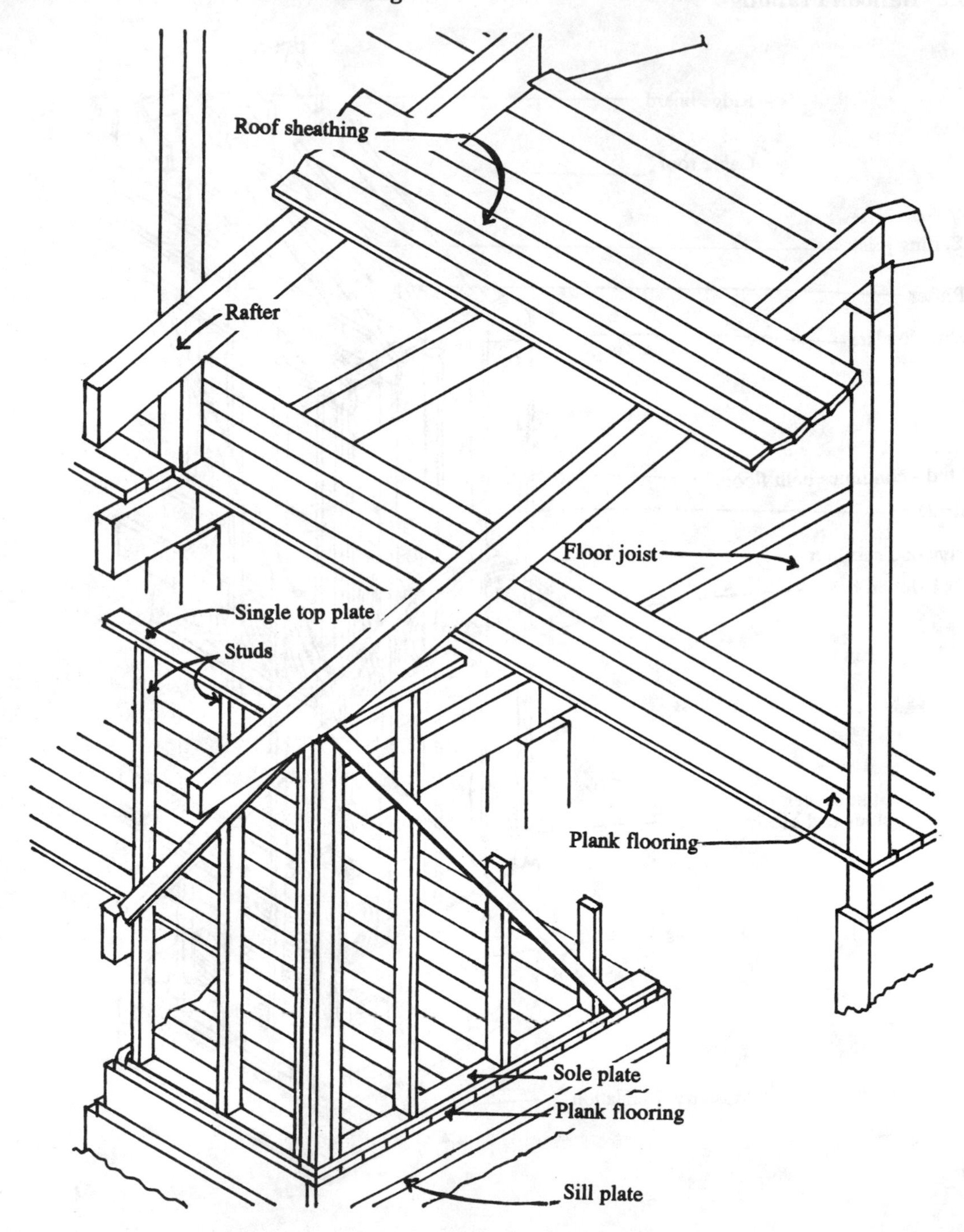

popular in contemporary styling, where exposed beam ceilings are desired.

Exterior Wall Framing

Until recently, when the cost of energy became so important, practically all wood frame houses were built with 2×4 studs placed 16″ on centers (o.c.), covered on the outside with an impregnated fiberboard sheathing with little insulating value and covered on the inside with drywall or plaster. In colder areas a 3½″ insulating batt was placed in the wall. Today, however, in many areas of the country, the insulating value of this type wall is inadequate.

Additional insulation can be obtained in two ways. First is the use of the 2×4 stud wall 16″ o.c. but replacing the impregnated sheathing with an insulating sheathing of polyurethane or similar product. Hang 3½″ insulating batts between the studs. For even greater insulation, use 2×6 studs installed 24″ o.c., the polyurethane sheathing and 5½″ insulating batts between the studs. Because fewer studs are used, the cost of the lumber remains about the same, but the 5½″ batts will cost more than the 3½″ batts.

If the area in which you are building warrants the greater insulation, then by all means go ahead and use it. On the other hand, if it is not needed there are advantages in staying with the 2×4 wall in that it reduces window and door installation problems. All other dimensions remaining the same, the use of the 6″ stud wall reduces the interior dimensions of the rooms of the house. This amounts to about 17 square feet of living space in a 2,000-square-foot house.

Appendix F, "The Superinsulated House," illustrates special framing, insulating and sealing techniques for houses built in very cold climates.

Long-Span Lumber

Figure 6.4 illustrates specially manufactured pieces of lumber for long spans not shown in previous drawings.

All are well suited for long spans as floor, ceiling and roof joists. They are becoming more popular in the building business and are easy to install, generally eliminate squeaking floors and are stronger than conventional lumber. When used as ceiling joists in a two-car garage, they can provide sufficient strength so that the usual girder with its pesky pipe column in the middle of the garage is not necessary.

FIGURE 6.4 Long-Span Lumber

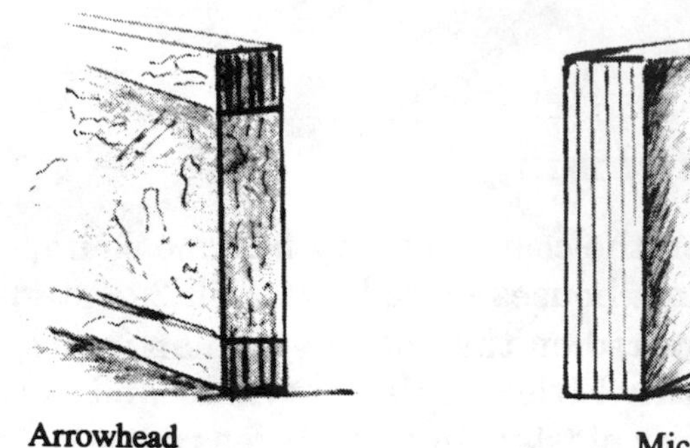
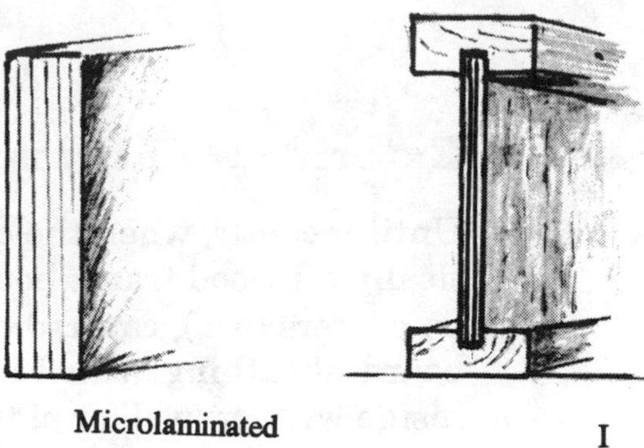

Tips on Good Framing

Glue the subfloor to the floor joists in addition to nailing or stapling. This provides a much sturdier floor system for little additional cost. Find out what size of glue cartridge the framing crew uses before buying the adhesive. Two sizes of caulk/glue gun are in general use.

If you are installing a plywood subfloor, leave a ⅛″ gap between the sheets. Plywood absorbs moisture and will buckle if it does not have room to expand. It also tends to delaminate when it gets wet. These delaminated areas must be replaced when the house is made weathertight. To eliminate this problem, use wafer board instead of plywood. It is waterproof and thus not affected by rain or other moisture.

Wafer board can also be used for roof sheathing, but it is slippery and may cause problems for the workers if the roof is steep.

Run two parallel strips of caulk between the top of the subfloor and the bottom plate on all exterior walls. This will reduce the loss of heat by air infiltration. (See Chapter 11, "Air Infiltration.")

Check with the framing crew on the best size of floor joist before ordering the lumber. The size of the joists required by the code is perfectly safe, but at the outer limits, the floor may be springy and have more bounce than you would like. Assume that the local code permits a 2×10 floor joist installed 16″ o.c. to cover a span of 15′ with the type of lumber you are using. This will be a safe floor but it will be bouncy. To stiffen the floor, you have these options:

- Use the manufactured long span lumber discussed above.
- Install the 2×10 joists 12″ o.c.
- Go to 2×12 joists.
- Install additional piers and girders in the foundation system to reduce the span to about 8′. Change the size of the joists to 2×8.

FIGURE 6.5 Crowning Joists

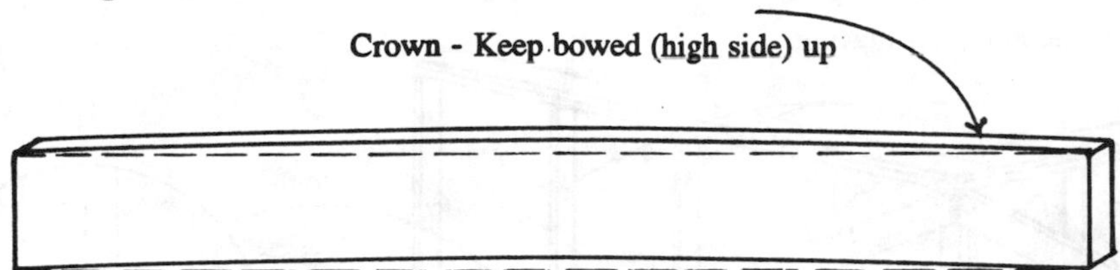

Consider the cost of these choices (or combination of choices) and select the one most advantageous for you.

Make certain that the framing crew "crowns" the floor and ceiling joists and the roof rafters (see Figure 6.5). Almost all lumber of any length has a crown in it.

The lumber should be installed with the high side up. Most framing crews will do this without being told, but it is worth checking.

Wood or metal bridging or blocking should be used to stabilize joists and rafters and to properly align them so that the subfloor or roof sheathing has a level base on which to lie (see Figure 6.6).

Metal bridging is easier to install than wood bridging. Blocking is the most solid and can be used to align a stubborn floor joist and hold it firmly in place before the subfloor is applied. Because the nailing of the bottom end of the bridging is not done until the house is more or less complete, this method will not hold an out-of-line joist or rafter in place as well as blocking. Either system should be applied between each joist and rafter at intervals of not more than 8′.

The most effective corner bracing is the diagonal brace "let into" the studs, as shown Figure 6.7. The braces can be placed inside or

FIGURE 6.6 Bridging and Blocking

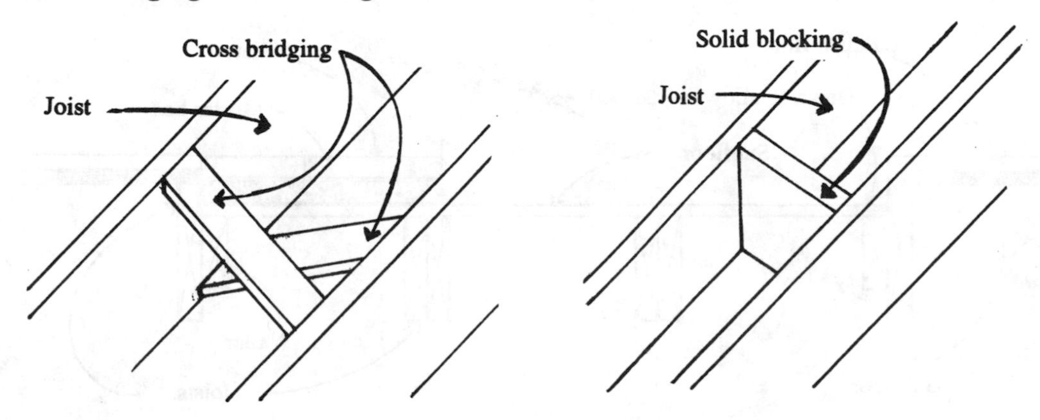

FIGURE 6.7 Let-in Corner Bracing

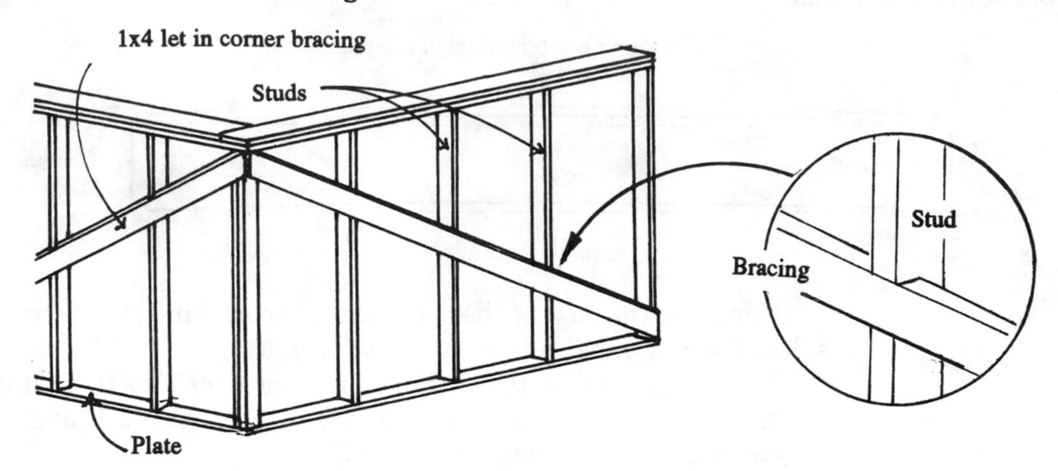

outside the studs. This type of bracing should be used if the sheathing is polyurethane or polystyrene, since either material has very little strength. Corner bracing of this type should be applied to all exterior corners and at intersections of exterior walls with interior walls. Sheets of plywood used as sheathing provide corner bracing also, but plywood has much less insulating value than polyurethane.

If your plans call for the use of slate, stone, earthstone or ceramic flooring in areas other than bathrooms, the subfloor must be dropped to allow additional space for setting of the flooring in a bed of cement.

As shown in Figure 6.8, this can be achieved by cutting out the top portion of the joist and installing the subfloor on the top of the modified joists. Check with the tile setter to determine how much

FIGURE 6.8 Dropped-Floor Framing

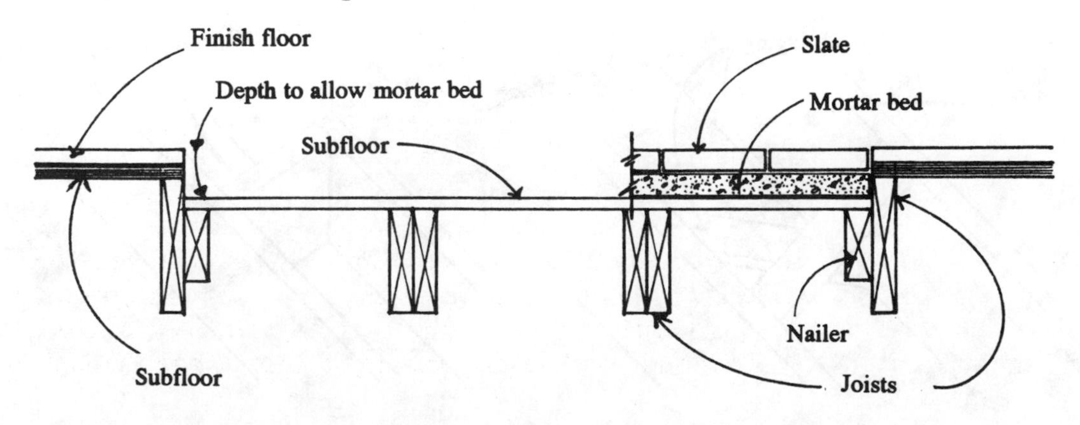

FIGURE 6.9 Drywall Nailers

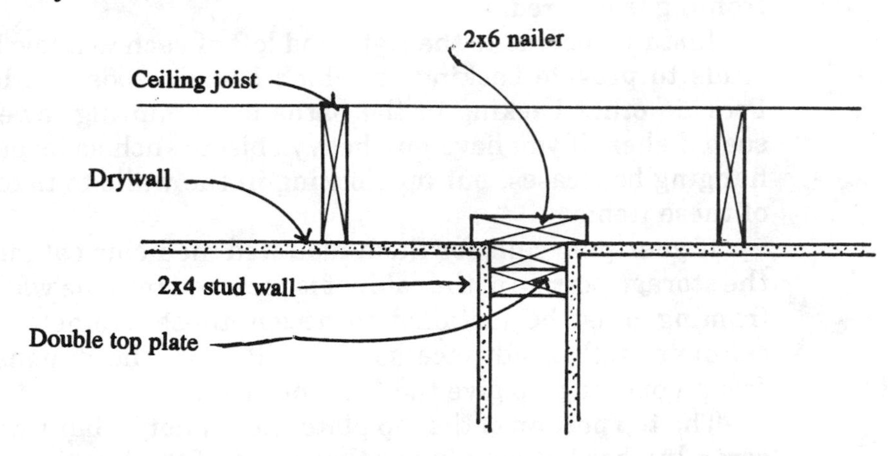

room is needed to have the finished floor even with the adjacent flooring.

In bathrooms, the ceramic tile is usually laid on top of the subfloor without lowering it. A marble threshold is set between the tile and the adjacent floor. The tile floor will be about an inch above the wood floor or carpet next to it.

Double the floor joists under the load-bearing interior walls. This measure will prevent sagging of the floor system.

Check that the framing crew has remembered to install nailers to which the drywall or plaster lathing will be attached (see Figure 6.9). At those junctures of ceiling and wall where the wall is at right angles to the joists, the joists provide the nailers on which to nail the drywall. On the other junctures, however, where the ceiling joists are parallel to the walls, there is nothing to attach the drywall to unless it happens that a ceiling joist lies at that point. If not, this is where you must provide the nailers.

Specify that exterior wall sheathing will be applied by hand nailing only. The polyurethane material is very easily damaged and torn by powered nailing systems.

A good framing crew will make accurate saw cuts. In particular, the cuts at the juncture of the roof rafters and the ridge board, those for the headers of the doors and windows and those for all intersections should be tight. If these cuts show large, obvious gaps, the workmanship is sloppy and could weaken the framing.

Use a long straight edge of 6′ or more to check the alignment of the studs and ceiling joists to find those that bulge in or out excessively. Mark them and show them to the crew chief, who should have them corrected. An out-of-line wall may cause problems in the work to follow such as hanging the drywall and, particularly, installing kitchen cabinets. It is relatively easy to make corrections before the framing is covered.

Install blocking to the right and left of each window between the studs to provide backing on which curtain rods can be mounted. Provide other backing in the baths for mounting towel racks and soap dishes. If you have any heavy objects such as large pictures or hanging bookcases, put up blocking in the walls to take the weight of these items.

If your plan calls for flush-mounted medicine cabinets in which the storage portion of the cabinet is recessed into the wall, additional framing must be installed to accept these cabinets. Select your cabinets well in advance so you will have the dimensions of the "rough opening" to give the framing crew.

The top portion of the top plate should not be built with all of the scrap lumber left over from other parts of the framing. The lengths of the pieces of the top portion of the top plate should be the same as the bottom portion of the top plate, but stagger the joints. Both the bottom and top runs perform an important structural function.

The size of the fiberglass tub-shower combination requires that it be installed during the framing. Arrange with the plumber to set it when the framing crew is ready.

Roof Framing

There are three generally used methods for providing support for the roof. Figure 6.10 shows the stick-built system using rafters, ridge board, ceiling joists and collar beams assembled on the job.

The second method is the prefabricated truss system in which the trusses are made to your specifications by a fabricator specializing in this work (see Figure 6.11).

The trusses are built using lower dimensional lumber than the stick-built system, but despite this reduction and the typical installation of the trusses at 24″ o.c., the truss design provides adequate strength to the roof. Savings in framing costs can usually be made using trusses because of reduced costs for labor and materials. They are normally designed to span the entire distance between opposite exterior walls. This arrangement provides more flexibility in interior-wall location since only the exterior walls are load bearing.

The truss does have one disadvantage in that the diagonals used in its design drastically reduce the usable attic space.

If your plans call for several dormer windows, the stick design will be the better choice, because it is much easier to tailor the roof system to accept the dormer framing.

If the plywood roof sheathing is 3/8″ on a truss system 24″ o.c., use metal clips (made for this purpose) between the sheets of ply-

FIGURE 6.10 Roof Framing: Stick-Built Construction

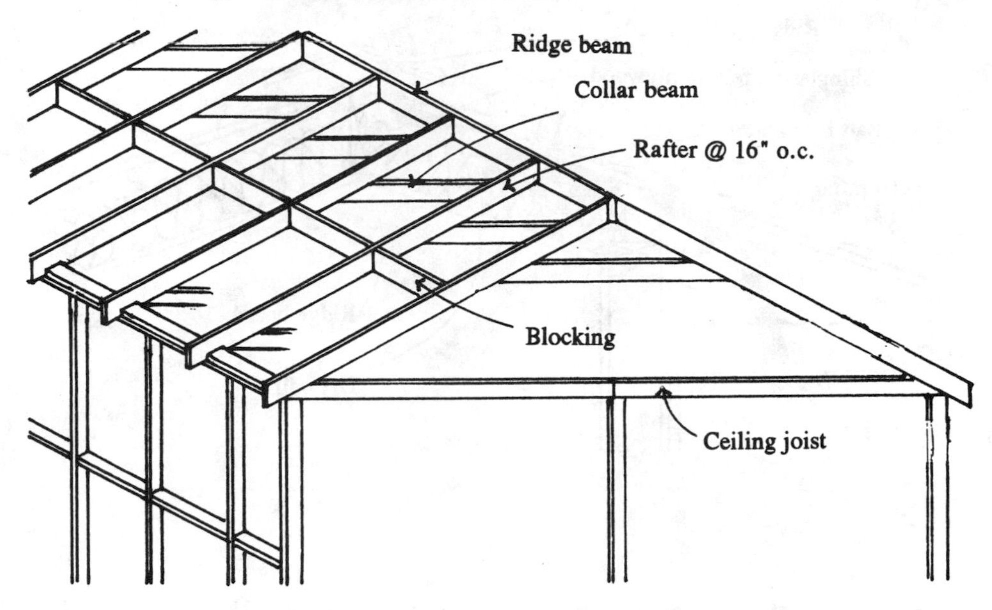

wood to reduce the potential sag of the sheathing between the trusses. Greater thicknesses of sheathing do not require the clips.

The third type of roof construction is the cathedral ceiling, which can be built of conventional lumber without exposed beams, as illustrated in Figure 6.12, or of lumber decking with large timber beams, as seen in Figure 6.13, or a combination of the two.

It is very important that the space between the ceiling and the roof be properly ventilated. With standard roof construction, either

FIGURE 6.11 Roof Framing: Wood-Truss Construction

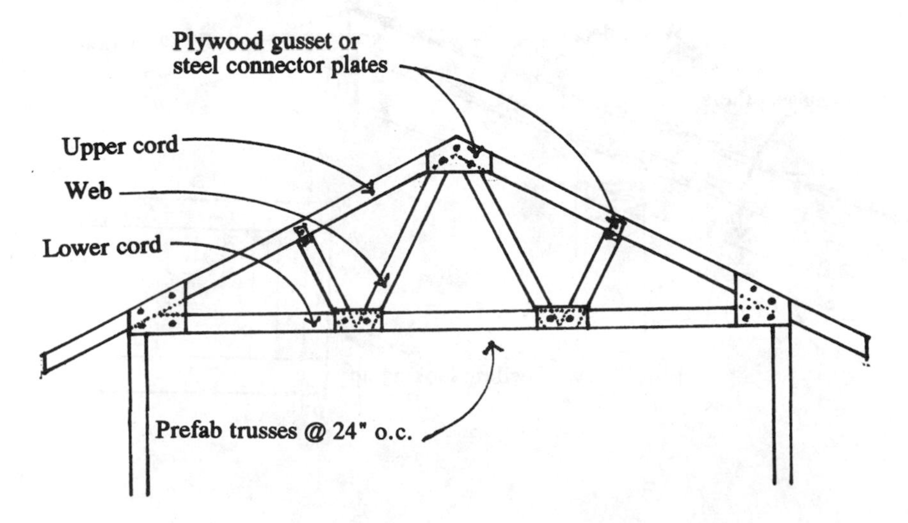

FIGURE 6.12 Roof Framing: Cathedral-Ceiling Construction

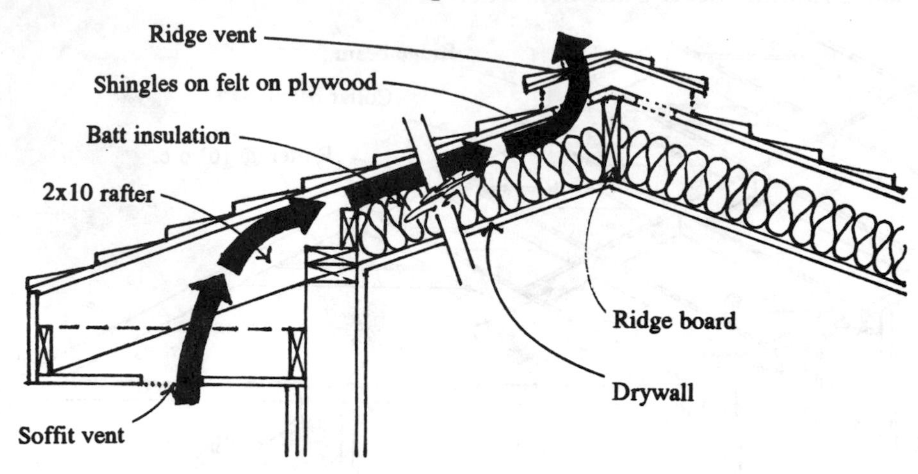

stick built or trussed, the attic is usually ventilated by the flow of air through openings in the soffit into the attic and then out through the vents in the gable.

With a cathedral ceiling, the ventilation is provided by a flow of air through the soffit over the ceiling insulation out through a ridge vent.

In a cathedral ceiling where the bottom side of the wood decking (used in lieu of sheathing) is the exposed ceiling, Figure 6.13, there is no air space requiring ventilation. The insulation, a rigid type, is applied above the decking with shingles on top.

FIGURE 6.13 Cathedral Ceiling Alternative

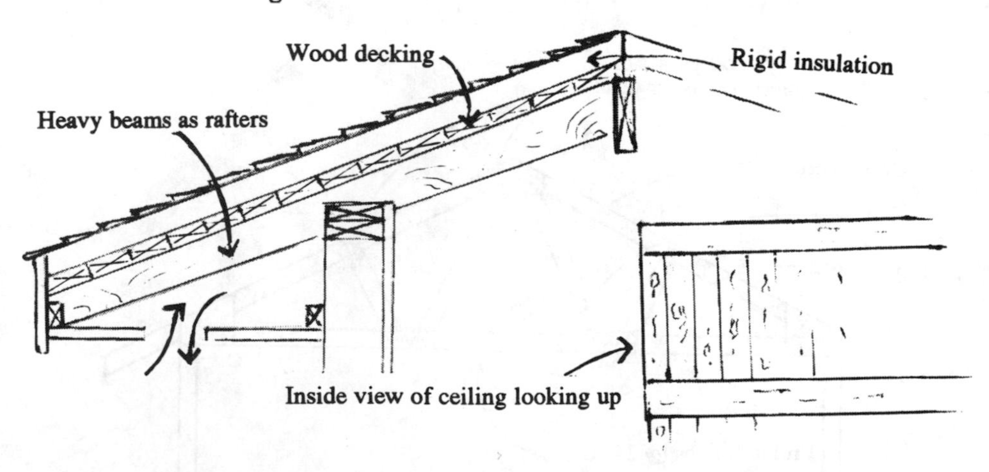

If you plan to install an attic ventilating fan, make sure the size of the gable end vents or ridge ventilation is sufficient to permit the passage of the amount of air required for the fan to operate efficiently (see Figure 6.14).

Foam-Panel Framing

A new method of framing a house has recently appeared and its use continues to grow. This new system uses factory-built solid panels in lieu of stud construction for exterior walls and for application on top of the roof joists or as ceilings.

These panels range in depth from 3½″ (R–14) wall panels to 11½″ (R–44) roof panels. The usual size is 4×8, but larger and smaller panels can be built. Note, see Figure 6.15, that the panels consist of wafer boards glued to a solid foam core with 1½″ lumber as top and bottom plates (4″ wide or larger to accommodate various foam thickness.) These panels are glued and/or nailed together and to the subfloor to form the exterior walls. Roof panels with similar construction constitute the roof structure for cathedral ceiling design or the ceiling structure of the top floor in houses where attic space exists under the roof. Interior walls are conventional stick built construction.

These panels are factory built at plants located mostly in the northern and eastern areas of the United States and in Canada, but they have dealers throughout the entire country.

FIGURE 6.14 Attic Ventilation: Gable Roof

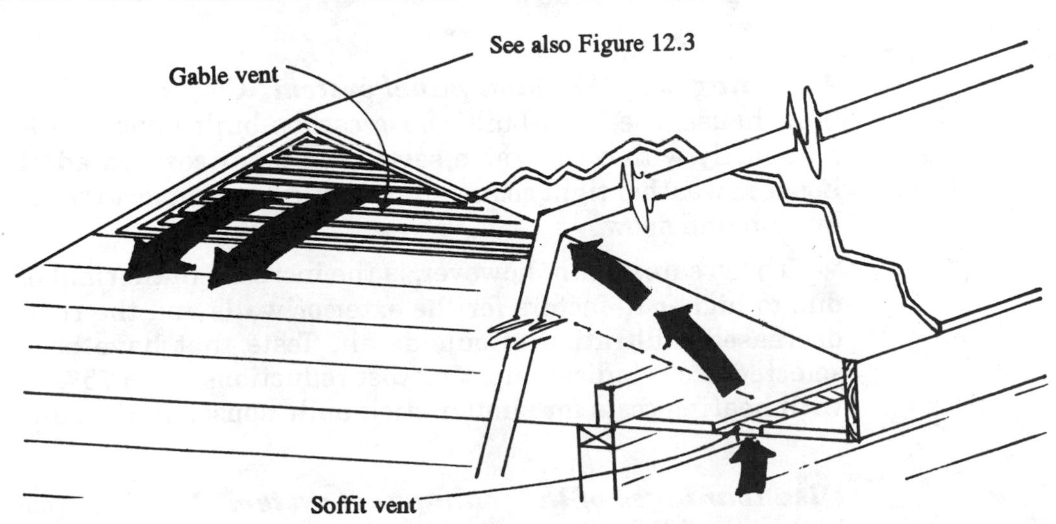

FIGURE 6.15 Typical Solid-Foam Panels

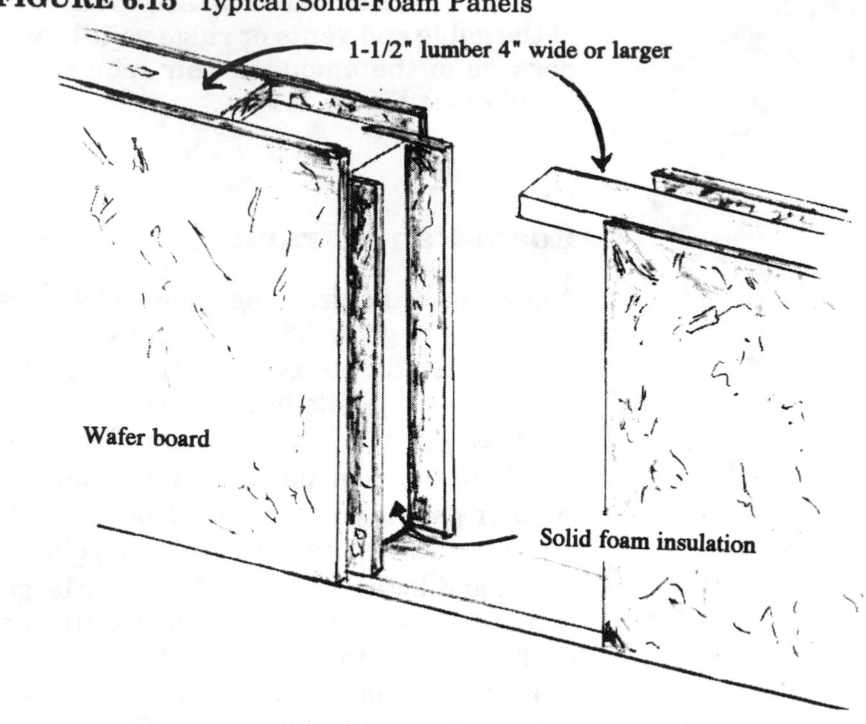

To enable the factory to construct the panels properly, complete plans for the house must be submitted including desired thickness of panels, the window and door cut out locations and sufficient details of electrical, plumbing and HVAC plans so that cuts for cable, pipes and heat ducts can be made. As an alternative, these panel cuts can also be made on the job.

Advantages of the foam-panel system Compared with the stick-built house, the foam-built house can be built much more quickly and easily with comparable savings in labor cost. In addition, the house is weather tight sooner, thus reducing damage to the structure by rain and snow.

The greatest gain, however, is the increase in thermal efficiency due to higher R–factors for the exterior walls and the roof and the decreased infiltration of outside air. Tests that have been run in selected areas indicate heating cost reductions up to 75% compared with heating costs for similar stick-built houses in the same area.

Disadvantages of the foam-panel system Material costs are higher than those for the standard stick-built house. Some or all of these costs can be recovered, however, by the reduced construction time and the substantial saving in heating costs.

Paneling cutting at the job site to accommodate pipe, cable and duct runs is difficult and increases labor costs but in complex house design it may be the best solution to reduce inaccuracies. Another solution, particularly for pipes and heating ducts, is to install additional wall framing inside the exterior panel walls and to run these pipes and ducts through the space provided. In any case, it is good building practice to run pipes and heating ducts through interior walls as much as possible to avoid heat loss if they are installed in exterior walls.

Once the structure has been completed, changes to the electrical, plumbing and heating systems will be more difficult to execute than with a stick-built house.

The foam-panel framing system continues to become more popular, although not all of its problems have been completely solved. Certainly this system should be investigated. Discuss the system with those who have built a house of this type before you make a decision about what kind of house you want to build.

Exterior Wood Trim

Figure 6.16 illustrates the various parts of the exterior trim or cornice work. Familiarize yourself with the names of these parts and where each is applied.

Trim material is usually 1″ nominal stock (¾″ actual) and the wood is superior in finish and quality to that used in framing because it is exposed to view. Among the better choices for trim material are spruce, SYP grade "C" or better, redwood and fir. Select one that fits your budget and satisfies your taste. The cost per board foot will be more than the framing lumber. See Chapter 8, "Exterior Finishes," for a discussion of all types of siding.

All exterior wood exposed to the weather, such as exterior trim and siding, should be back primed, that is, sealed on the nonexposed side before installation to reduce cupping, other distortion and exterior paint peeling. One coat of the prime paint you will use on the exposed side will do the job.

Costing

In most cases, the framing crew will furnish the labor and tools. If they use a powered nail-gun system, they will also furnish the nails.

FIGURE 6.16 Exterior Trim

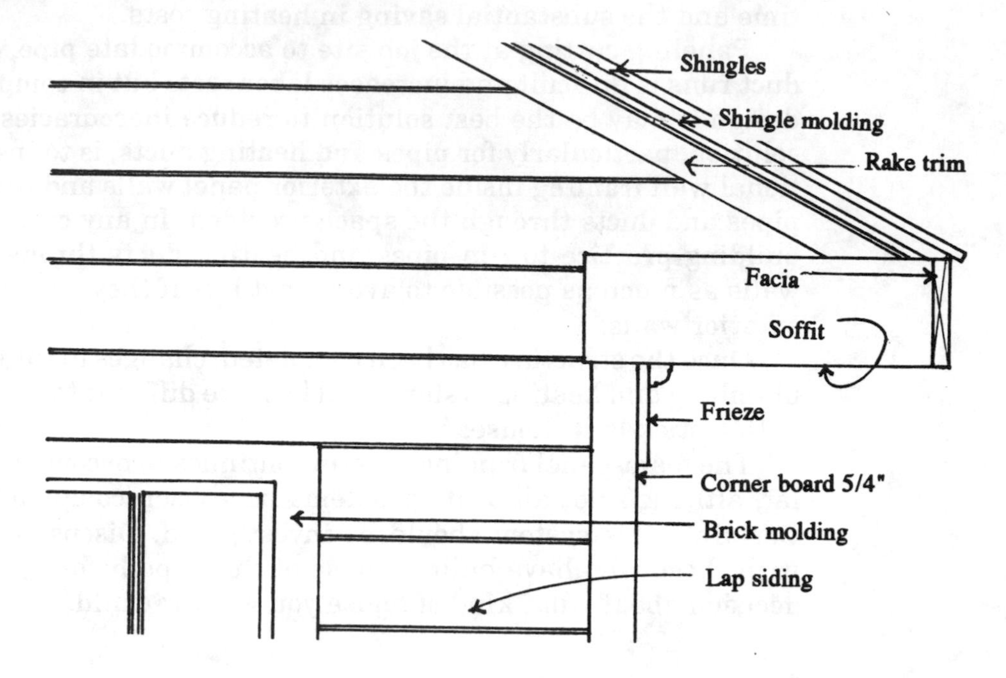

The cost of the labor for framing and exterior trim, therefore, will be provided by the bids of the carpentry contractors.

Most suppliers will make the lumber take-off from your plans to include the framing material, exterior trim, siding, doors, windows, interior trim and the rough hardware (nails, joist hangers and other metal pieces used to put the wood together). Have them break down the cost into categories so that you can take advantage of the best price in each case. In other words, you may want to order the framing lumber from one supplier, the doors and windows from another and the trim material and roof trusses from still others.

Lumber is usually priced as so much per thousand board feet. A board foot is theoretically a board 12″ wide, 12″ long and 1″ thick with all dimensions being nominal. To determine the amount of board feet in lumber, you must translate its dimensions (using nominal, not actual) into board feet. For example, a 2×4 that is 1′ long is equivalent to a one inch board 8″ wide and 12″ long; thus each linear foot of 2×4 equals .67 board feet. Following the same procedure, one linear foot of 2×6 equals one board foot, one linear foot of 2×8 equals 1.33 board feet, one linear foot of 2×10 equals 1.67 board feet and one linear foot of 2×12 is equal to two board feet.

Applying these factors to the requirements of a particular plan will give you the total board feet. For example, if the floor joist requirement is 24 pieces of 14′ long 2×10s, the total board footage for floor joists would be 1.67×14×24 = 562 board feet. If the price of

SYP is $624 per thousand board feet, the cost of the floor joists would be 624×562 divided by 1,000 or $351.

Suppliers will do the pricing for you from your plans, but it is a good idea to know how to do it yourself so you can compute the cost of additional lumber needed during the building process.

Management

Electric power will be needed for saws and other tools used by the framing crew. A temporary electrical hookup should be provided by the electrician, which is then connected to the local power source by the power company. In an emergency, portable generators can be used, but power from the temporary hookup is much preferred.

Get together with the framing crew chief to plan where the lumber and other materials are to be placed on site by the supplier. If you have a flat, open lot this will not be a problem, but if your lot is wooded with little open space or if the supplier's trucks cannot move to the interior of the lot because of soft ground, you may have a storage problem. If so, the solution is to have the supplies delivered in phases: first, the floor system, then the wall material and finally the roof material. In the event that more than normal hand carrying of the lumber is required of the framing crew, you should expect to pay more for the framing labor, unless this problem was included in the information you provided to all bidders.

Most lumber suppliers use flat-bed dump trucks for delivery purposes. Expect to have your lumber dumped into one or two large piles. These piles will need to be sorted and restacked before framing begins. Have an understanding with your framing crew that this restacking will be their job.

The crew chief should know the rough openings of the windows and the size of the exterior doors you have ordered before the wall framing begins. If they are not in your plans, get the information from the door and window supplier.

If your framing crew uses conventional nails, include their cost in the framing materials. It is much cheaper to buy nails in 50-pound boxes (less than half price), so buy in these quantities if close to your needs.

Visit the job daily to keep up with the work and to see that the crew does not run out of materials. Often the crew chief will forget to tell you of pending shortages, so ask each day about supplies in each category. Be prepared to go after items that are needed immediately and can be carried in whatever car or truck you are using.

Before accepting the materials bid, find out from the supplier what delivery service you can expect. You will be wise in using a supplier who offers daily deliveries rather than one who delivers only twice a week, even if the former is somewhat higher in price. In the ideal arrangement, you would check on the job in the afternoon and then phone in any order to the supplier that day or before 7:30 A.M. for delivery by the first truck leaving the supplier's yard.

Windows and doors should be delivered to the site just before they are to be installed. If you have them sent out too early, there is more danger of breakage and absorption of excessive moisture by the wood frame. Order interior doors for delivery along with the rest of the interior trim material, after the house is weather tight and the drywall or plaster has been finished.

If your house has full-height exterior brick veneer walls in whole or in part, it is easier for the exterior trim work to be done after the brick has been laid. Line up the brick mason at the proper time, but do not have the work start until the framing has reached the point where the brick work can be completed without interruption.

Use aluminum or hot-dipped galvanized nails for all woodwork exposed to the weather, even if it is to be painted. Electroplated galvanized nails do not hold up as long as the other choices, and they tend to rust through the paint or stain.

The method of placing the drywall or plaster lathing inside the house should be considered during the framing. Those heavy, awkward, dry wall boards will probably be delivered by a special truck with a boom designed to handle this material. In most cases the best solution is to pass the drywall through a window opening in which the window has not yet been installed or which can be easily removed. Another method is to leave off one or two sheets of sheathing and pass the drywall through the opening. Discuss this problem with your crew chief and drywall supplier.

7

Windows and Doors

In these days of high energy costs, the selection of windows and doors assumes much more importance than it did a few years ago. In making your selection, consider insulating and anti–air infiltration properties, ease of operation, maintenance, ease of cleaning, style to suit your architecture and price to fit your pocketbook, in just about that order of priority. Do not buy cheap windows that will cause a loss of any savings through increases in heating and cooling costs and maintenance.

Windows

The following types of windows, illustrated in Figure 7.1, are those readily available on the market:

- *Casement* Operates like a door in that it is hinged on one side and rotates on a vertical axis by turning a crank. Along with the awning and hopper windows, it offers the best seal against air infiltration. Depending upon the particular design, interior and exterior surfaces can be cleaned from inside the house.
- *Double hung* The popular window found in most houses. Its two sections move in a vertical direction in channels and are held in place by springs or friction. The single-hung window is a variation of this style, in which only the bottom of the two sections moves.

FIGURE 7.1 Types of Windows

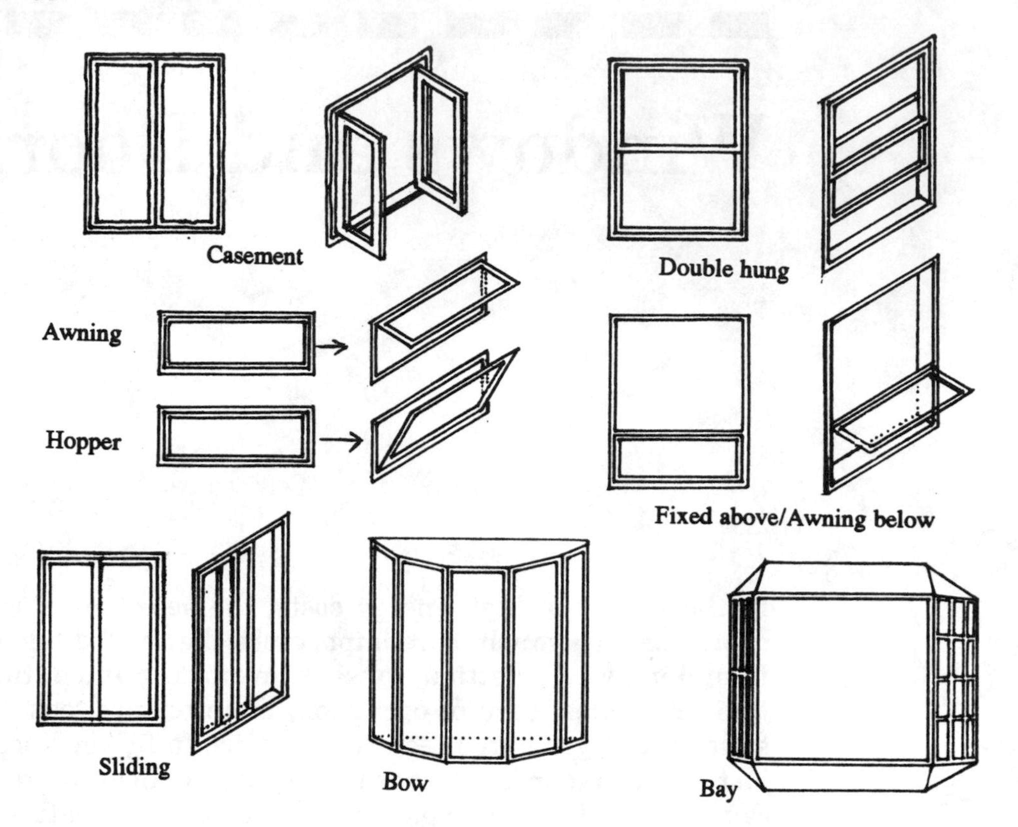

- *Awning* Operates very much like the casement except that the hinged rotation (to the outside) is horizontal. The awning window is often used in combination with a fixed window above or below.
- *Hopper* Operates like the awning window except that the hinge is at the bottom with rotation to the inside.
- *Sliding window* Slides horizontally in a channel in the frame.
- *Fixed* Glass mounted in a frame with no movement. It can be ordered in a variety of sizes and shapes.
- *Bow* A window made up of several elements in the shape of an arc extending out from the exterior walls of the house. All sections are usually fixed.
- *Bay* Similar to the bow window except that it has two side sections that are angular instead of curved and a straight center section. Bay windows are sometimes composed of combinations such as casement or double hung windows on the sides and a fixed center section.

Insulating Glass

Insulating glass exists in two forms. One is a special double sheet of glass separated by an air space with the glass edges welded together to form an airtight center space much like the liner for a thermos bottle. The other consists of two sheets of glass held in the frame with an insulating air space in between. For most areas of the country, the windows you select to install in your house should use one of these two systems. The additional expense is well worth it, and you should save enough in fuel costs to more than make up for this expense in a few years and have a more comfortable house in the meantime.

Those who live in very cold climates should consider using windows with triple glazing, that is, three layers of glass with two separate air pockets in between.

Glass itself is a very poor insulator and will act as a high heat loss area unless air spaces are interposed to block the ready escape of heat. Even then the insulating qualities of the double- and triple-insulated glass windows and doors will provide only ⅛ to ⅕ the insulating value of a 2×4 wall with 3½″ of insulating batt and a 1″ thick polyurethane or polystyrene sheathing. On the other hand, windows installed to take maximum advantage of the sun to provide heat can be a definite asset. See Chapter 13, "Solar Heating."

Types of Glass

- *Tempered glass* should be used in doors, glass panels adjacent to doors and other areas where it is likely that people, especially children, might fall against the glass and break it. Most codes require the use of tempered glass in wall areas that are within 4′ of a door. When broken, it crumbles into more or less harmless granules.

  Tempered glass cannot be cut, so be certain that your measurements are correct before ordering. Allow about ⅛″ on each side of the frame for expansion. Standard sizes cost less and are usually readily available from stock.
- *Annealed float glass* (formerly called plate glass) is used in house windows. It breaks easily into pieces that are usually very sharp. This glass is largely being replaced by Low E glass, discussed below.
- *Tinted glass,* usually annealed float, is designed to reduce the passage of solar heat. It is particularly useful in areas such as the seashore where glare is obnoxious and in the Southeast, Southwest and similar environments.

- *Reflective glass* serves a similar function as tinted glass but is more effective. It reduces solar heat passage to about 50% of that of annealed float glass, whereas tinted glass offers only a 25% reduction. This type of glass was designed primarily for commercial buildings with large window areas to reduce heat loss to the outdoors in the winter and to reduce heat gain in the summer. From the outside, the windows look like mirrors, which most homeowners will not want.
- *Low E glass* (low emissivity) is used in dual glazed windows with a low E coating on the inside to reduce heat loss in the winter and heat gain in the summer. It is an excellent choice for houses.

Window Frames

Windows may be selected with the following choices of frames:

- *Wood* of excellent quality woods is frequently used. Because frames must be stained or painted, maintenance costs are higher than those for other choices. The color of the finish depends upon the selection of the paint or stain.
- *Wood clad with vinyl* makes a very fine window with a long life. The wood frame is completely encased in a thick layer of tough vinyl. Because no painting or staining is required, the maintenance on this window is very low. The choice of color is usually limited to white, brown or bronze.
- *Wood clad with aluminum* is another fine choice with almost no maintenance. In this type, the wood frame is encased in a layer of aluminum, which is coated with baked-on, factory-applied paint. The choice of color is usually limited to white, brown or bronze.
- *Aluminum*-framed windows are available in natural finish or several selections of anodized coloring. This is a long life frame that is inexpensive compared with other materials. Aluminum is an excellent conductor of heat, though, and will permit high heat losses from within the house through the window frame. It will cause excessive moisture condensation on the interior portion of the frame when there is a large difference in temperature between the outside and the inside. For these reasons, if you buy aluminum-framed windows, be sure to select a brand that has a thermal break using an insulating material to separate the exterior and interior sections of the metal frame.
- *Steel*-framed windows are sometimes used in basements. They have the same disadvantages as the aluminum-framed win-

dow without the thermal break and, in addition, they must be kept painted to prevent rusting.

Screens

Screens should be included in the window specifications for the parts of the window that will be opened. Nylon is the most popular screen material. It will not corrode like copper and aluminum. When the windows are delivered to the construction site, remove the screens and store them in a safe place until they are needed. They will probably be damaged if allowed to remain on the job.

Screen doors should be provided for entry ways that will be habitually open during the warmer months.

Storm Windows

Properly installed storm windows provide additional insulation and present another barrier to air infiltration. They are difficult to operate, however, and it is preferable to obtain the additional insulation through dual or triple glazing and the reduction of air infiltration through efficient design and construction rather than the use of storm windows.

Skylights

Skylights are a form of fixed window mounted in the roof. They are particularly popular in contemporary styles and are useful in taking advantage of passive solar heating. Figure 7.2 illustrates the most commonly used skylight in house construction.

The preferred skylight is one made of plastic and double domed to provide insulation against heat loss through the roof. The type that is mounted on a job-built curb is the choice selection for home building, since it is less troublesome and provides greater protection against leakage. Have the skylight on the job before the framing for it is begun to ensure that the skylight curb and the roof framing fit. To prevent water leaks, follow the mounting instructions of the skylight manufacturer explicitly.

FIGURE 7.2 Skylight Installation

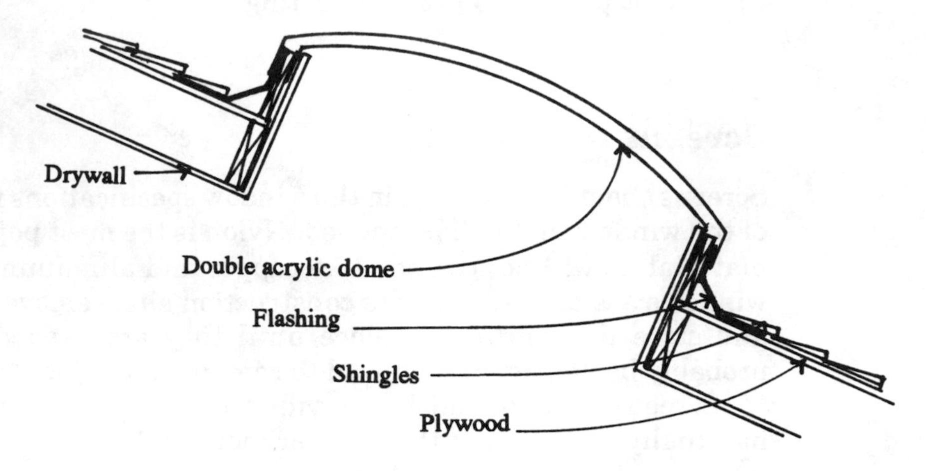

Exterior Doors

Like windows, exterior doors must perform a more important function in this era of high energy costs. They must effectively seal the opening against air infiltration, insulate against the loss of heat through the material and still provide easy movement through the wall. There is much more to door selection today than just appearance and style.

Wood Exterior Doors

Wood doors are made in a variety of styles to satisfy almost any taste. Figure 7.3 illustrates only a few of these styles.

Exterior doors are thick and heavy, which they should be, and can be selected from a variety of widths and with heights of 6′8″ (standard) and 7′0″. The two most popular styles are the flush door and the paneled door.

There are two types of flush doors: the solid core door made with particle board between the two outer surfaces and the solid lumber core door with solid lumber between the two outer faces. The latter is the much better and stronger door and the most expensive.

Most wood exterior doors are made of high-grade fir. The advantages of wood doors are their beauty, particularly if stained to show off the grain, and the vast array of choices. Their disadvantages are that they can warp, crack or shrink and cause sealing problems and

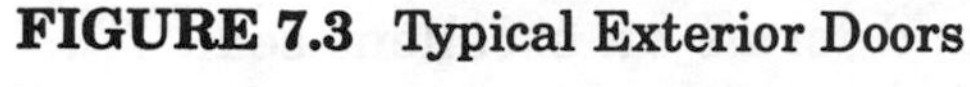

FIGURE 7.3 Typical Exterior Doors

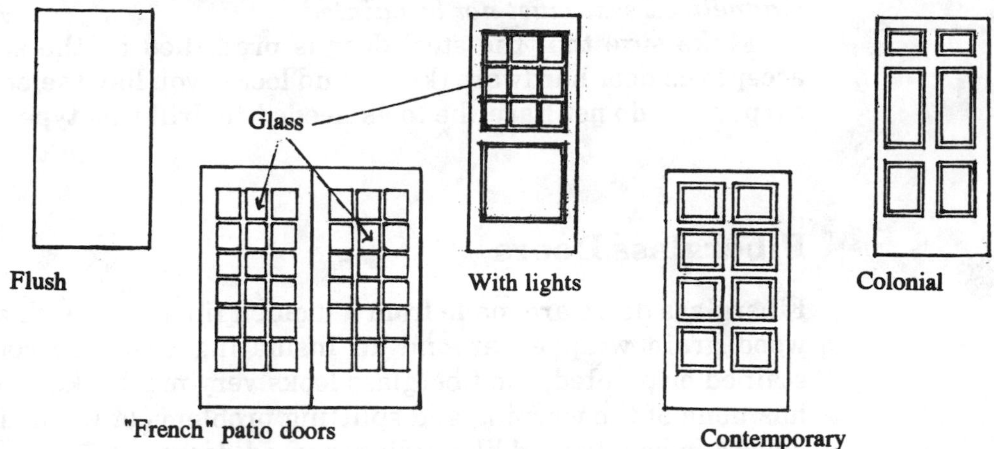

that other choices provide better insulation. Both paneled and flush doors are made with glass inserts that vary in number and shape.

Molded Exterior Doors

These doors have an advantage over the wood door in that they are less likely to warp, crack or shrink. They usually cannot be stained, however, and show little, if any, grain.

Steel Doors

Steel doors are made with a steel outer shell filled with insulating material. They are usually prehung, that is, they are supplied with their own frame, jamb and hinges and are either already assembled or the material is precut for easy assembly on the job. The steel door provides an excellent seal against air infiltration, is well insulated, is more difficult than other types to force open and is not affected by moisture. Its principal disadvantages are that it must be painted and usually the panel embossing is more shallow and not as handsome as that in a wood door. Plastic decorator panels that can be applied to the face of the door provide considerable variation in design.

Some steel door systems are made with magnetized weatherstripping (similar to that used on modern refrigerators), which clamps to the door when closed and forms an excellent seal. *This magnetized seal must not be painted.*

Make sure that the steel door is predrilled by the supplier to accept the door hardware (knobs and locks) you have selected. Most carpenters do not have the tools needed to drill this type of door.

Fiberglass Doors

Fiberglass doors are made from a molded fiber skin with an etched wood grain wrapped around an insulating urethane core. When stained or painted, the fiberglass looks very much like wood, but it has none of the warping and splitting problems of wood. The fiberglass can be trimmed like ordinary wood doors.

Patio Doors

Most patio doors consist of sliding and stationary glass panels with widths of 3′ or more per panel (see Figure 7.4). Their frames are available in the same choices as the windows discussed earlier and with single, dual or triple glazing.

Sealing to stop air infiltration through doors is just as important as it is with windows. The sliding patio door has the same disadvantage as the slider and the double-hung window. By the nature of its design, it is difficult to properly seal these doors against air infiltration. Even with higher-quality doors, the wear and tear caused by the door movement across the sealing material is much more prone to eventually cause leaks in the sealing than the conventional nonsliding door.

If large amounts of light and sunshine are desired (see Chapter 13, "Solar Heating"), they can be obtained by using single or double

FIGURE 7.4 Sliding Patio Doors

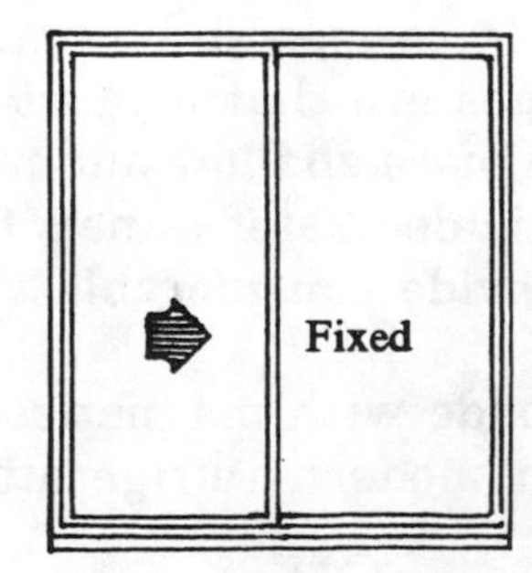

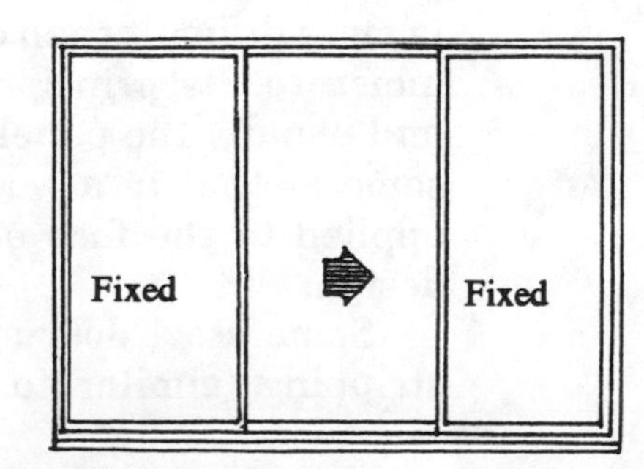

swinging doors along the lines of the French door or, even better, by a combination of one fixed panel with a swinging door. If passage through the door is not needed, then the whole system can be fixed and eliminate the air-infiltration problem.

Interior Doors

With interior doors, there is usually no problem involving insulation and air infiltration. These doors are not normally weatherstripped. As a matter of fact, if the heating system depends upon the free flow of air, interior doors should be undercut at least ½″ above the finished floor or carpet to permit this air flow. This does not apply in those rooms with both air supply and return outlets.

Wood Interior Doors

Wood interior doors can be purchased in various forms such as paneled, flush or louvered. The flush doors can be solid or hollow and made with finished surfaces of birch, mahogany, pine and hardboard. The hardboard cannot be stained, but it will take paint well.

The paneled and louvered doors are usually a species of quality kiln-dried western pine.

Louvered doors are available as full louvered, top and bottom louvered or partially louvered with either the top or the bottom paneled. They are particularly useful in locations such as closets where the free flow of air is desired. They may be required for rooms needing ventilation, which contain mechanical equipment, heaters for hot water and heating equipment.

Molded Doors

Interior doors molded of wood fibers are available primarily in the paneled form. They are much less expensive than the wood-paneled doors but will not take staining. They are hollow except at the points where the door hardware is to be applied and will not readily accept coat hooks or towel racks. They will not warp or split as wood doors occasionally do.

Prehung Doors

A prehung door is supplied already hinged to its frame and thus some of the trim labor is saved. Although it is more expensive than the non-prehung door, the savings in labor usually results in an overall cost reduction. Since prehung doors are supplied with part of the casing installed and part precut but not installed, you must select the type of door casing for the entire house before you order the doors.

Pocket Doors

A pocket door slides in and out of a pocket built into the wall framing. Like the folding door, it is very useful if there is little room to swing a standard door out of the way. On occasion the pocket door can cause problems if not carefully installed with quality materials. In order to provide the pocket, the size of the stud is substantially reduced, producing a frame where warpage is much more likely. This warpage can interfere with the operation of the pocket door. To minimize the probability of warpage occurring, use 2×6 studs in those walls in which a pocket door is to be installed.

Bifold Doors

The bifold door (see Figure 7.5) is made of two or more hinged sections. For small openings, it usually consists of two narrow panels hinged together to form one door. For large openings, bifold doors are made with wider panels, four narrow panels (forming two doors) or a combination of both. These doors are designed to be operated from one side only and, consequently, are best suited for

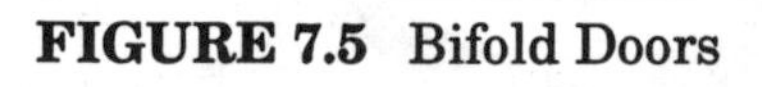

FIGURE 7.5 Bifold Doors

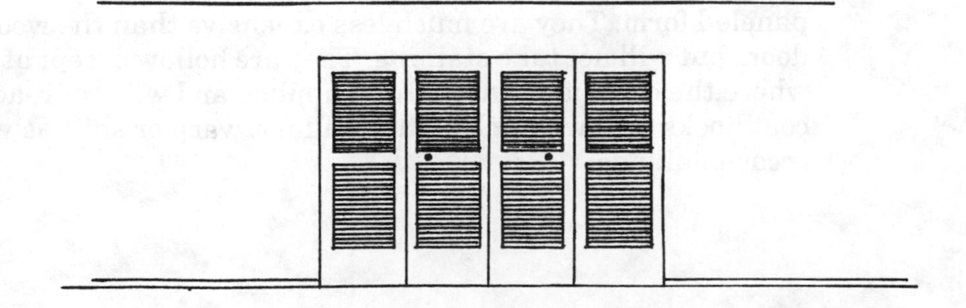

FIGURE 7.6 Closet Design

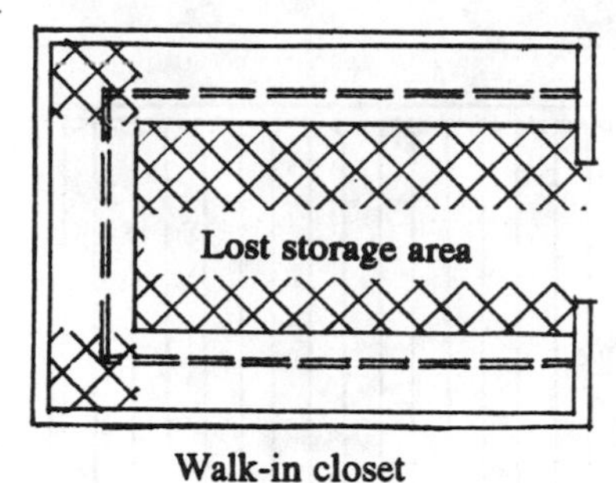

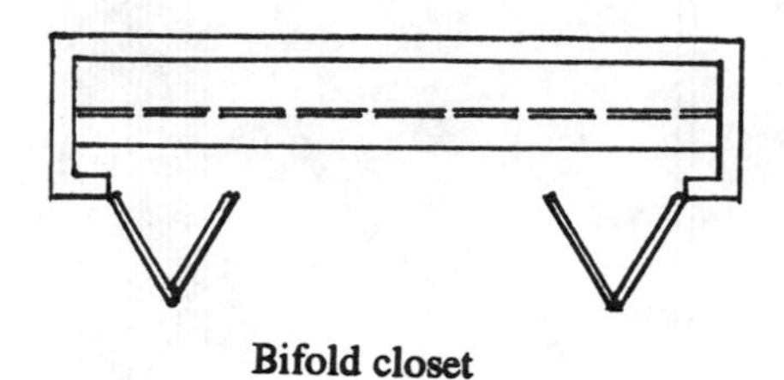

closets. They are most useful in providing wide doors to shallow closets, making the most use of the square footage available.

Figure 7.6 illustrates walk-in closet design. Note that the space used for the "walking in" cannot be used for storage. In the shallow closet with the bifold door, however, the user does not enter the closet, so that almost all of the space is available for storage. A disadvantage of the bifold-door closet, however, is that much of the wall space is lost to the door and the flexibility of the room furniture placement is reduced.

Folding Doors

The folding door, shown in Figure 7.7, can be made of wood, vinyl and many other materials. It is most useful if there is little room to swing a regular door.

Costing

The cost of the windows and doors will be provided by the supplier. The cost of the labor to install them should be included in the framing and interior trim bids. The cost of priming the exposed exterior wood should be included in the painter's bid. Make sure that this is understood before the bids are made. This work involves a separate trip and a few hours of labor for one or more painters. If you have selected windows and doors that do not have exposed wood, make this fact known to the painting contractors also.

FIGURE 7.7 Folding Door

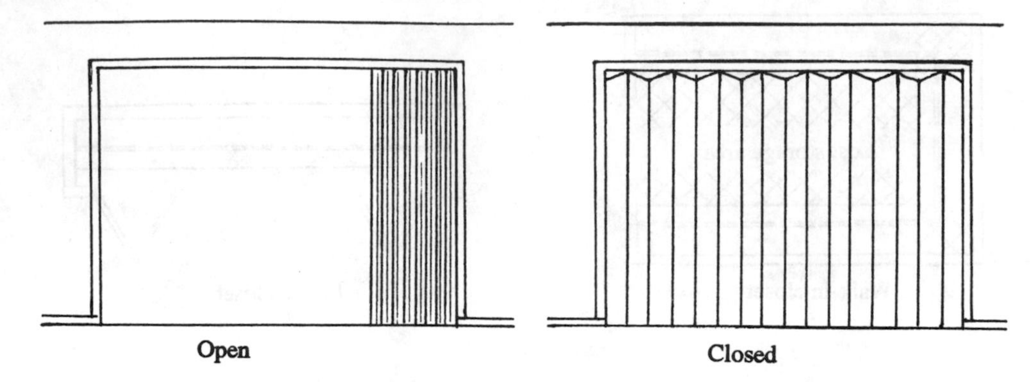

Management

- Order your doors and windows well ahead of time. Many types are held in stock by the suppliers, but if your choices include some that are not stocked, delivery can take several weeks.
- Get the rough openings of both windows and doors from your supplier and pass them to the framing crew. They will need these dimensions during the framing to cut headers of the correct size for the openings.
- If you have brick veneered exterior walls, you will need to know the window and door widths to order the proper size steel lintels for the brick.
- Among those items most changed from original plans are the types, sizes and locations of windows and doors. Be sure that the copies of plans available on the site have the correct placement of doors and windows and show the rough opening measurements for the actual doors and windows you have ordered. Do not request delivery of the windows and exterior doors to the site until just before they are to be installed. This reduces the exposure to theft, breakage and excess moisture absorption.
- As soon as they arrive on site, large glass areas should be clearly marked with an "X" of masking tape or other suitable material to reduce the chances that someone will walk through the glass or push a piece of lumber through it. Unfortunately, this type of accident happens too frequently.
- Collect the screens, muntins and uninstalled hardware (such as cranks for casement windows, but leave one on the job so that the carpenters can operate windows) and store them in a safe place away from the construction area to reduce the possibility of loss or damage.

- **Prime the exposed exterior wood shortly after (or even before) installation to protect this wood from the weather.**
- **After installation, check the operation of all of the windows and doors to ensure that they are not binding. If the operation is not smooth, correct the installation now. It may become a major problem to correct after the house has been finished.**
- **Check to make sure that the painters have painted both the tops and bottoms of wood exterior doors to prevent moisture from getting into the wood and damaging it. Many wood and wood-material door manufacturers will void their guarantee if these two areas have not been painted properly.**

8

Exterior Finishes

Almost any house structure can be altered to fit any of the available exterior finishes: brick, stone, wood, stucco, aluminum, vinyl or any other. In most cases, there is little or no requirement to change the plan. The principal exception is the plan based on wood or similar siding, which must be altered to accept brick or stone veneer.

To make the change to brick veneer from wood siding, you have two choices. You can move (1) the exterior wall inward about 5″ to allow space for the full brick veneer and thus retain the same outside dimensions for the exterior walls, or (2) keep the exterior load-bearing foundation walls the same but install the brick veneer outside these walls. The first choice should require no changes in the roof structure, particularly the overhang and the exterior trim, whereas the second choice may require changes in these areas to accommodate the wider dimension of the outer limits of the exterior walls. In the first choice you will lose some living space in the house—roughly equal to about 5″ multiplied by the length of the exterior walls of the living area.

If you choose to retain the exterior wall in its original position and move the brick veneer out, be sure to check the ceiling joists and roof rafters to see that their length will accommodate the change. The same caution applies to the dimensions of the roof trusses if your plans use them.

Brick and Stone

Brick Exteriors

Brick makes a very attractive exterior. It has a very long life and rarely requires any maintenance. It is, however, more expensive than wood, stucco, vinyl or aluminum. Although brick does not require painting, any wood trim of the house will require it. And the trim is relatively more expensive to paint than the siding because it takes more labor for similar square footages.

A silicone sealer should be applied to the chimney (if it is made of brick) to prevent water leakage into the house. If the brick selected is very porous, a double coat of sealer may be required on the exterior surface of all the brick.

Openings over doors and windows require some special support to hold the weight of the brick above the opening. There are three methods in general use, as illustrated in Figure 8.1:

- *Steel lintels* are the simplest and easiest to install. This type of lintel consists of a steel angle iron of the appropriate length with flanges about 3½″ wide and a thickness of about ¼″. The length should be sufficient to allow it to span the opening with not less than 8″ bearing on each of the adjacent supporting walls. For larger openings such as a double garage door, use a

FIGURE 8.1 Lintels and Arches

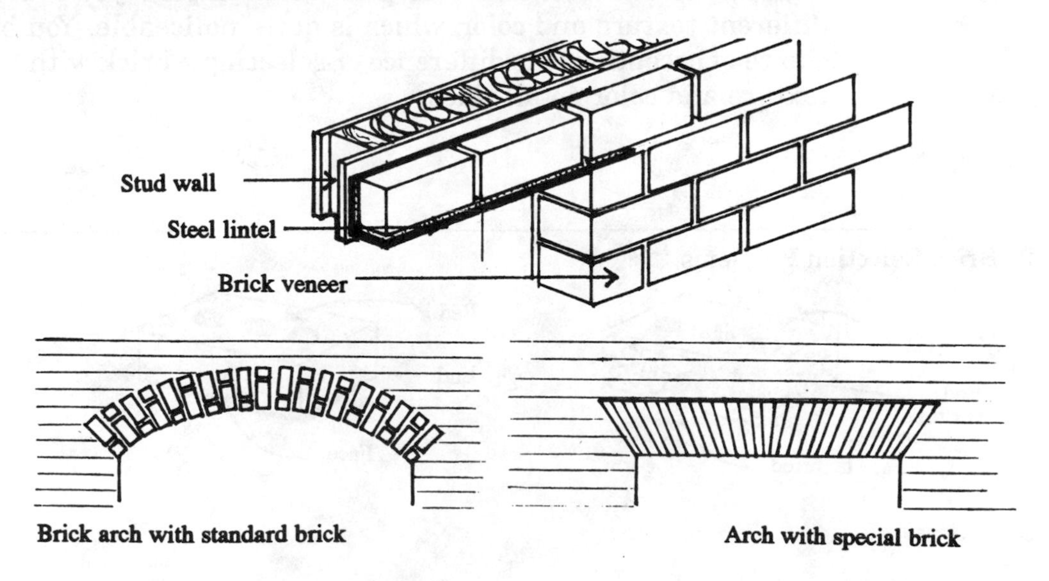

larger steel angle iron of about 5″×3½″×5⁄16″ thick. The wider flange of 5″ should be installed in the vertical position.

- *Curved brick arches* can also be used to span the opening. The brick used in this type of arch is the standard size and shape. It must be supported by a wood form until the mortar has set so that the design of the arch can carry the load. Note that the curve of this arch will require that the frame and trim of the door or window installed below it must be curved to match.
- *Flat brick arches* require expensive, specially cut bricks. The job should be done by an experienced mason, preferably one who has done this type of work before.

Brick Selection

To provide better adhesion with the mortar, much of the brick available is made with holes throughout.

These holes are concealed in a normal brick wall, but if your brick work requires that the bed of the brick be exposed (such as the end brick of a step), you will need to have some "solids," that is, brick without the holes (see Figure 8.2). Check with the supplier to find out how the manufacturer supplies these solids. Some include the solids as part of each packaged cube (approximately 450 brick). Others require that they be ordered separately. Your mason should be able to give you an idea of the quantities of solids needed.

Some brick is made with a textured facing that appears only on the surfaces exposed in a running or common bond. If part of your masonry work requires exposure of the large flat surface or bed (again, the end brick in a step is an example), you will have a different texture and color, which is quite noticeable. You have the choice of accepting this difference or selecting a brick with the same texture and color on all sides.

FIGURE 8.2 Brick Selection Problems

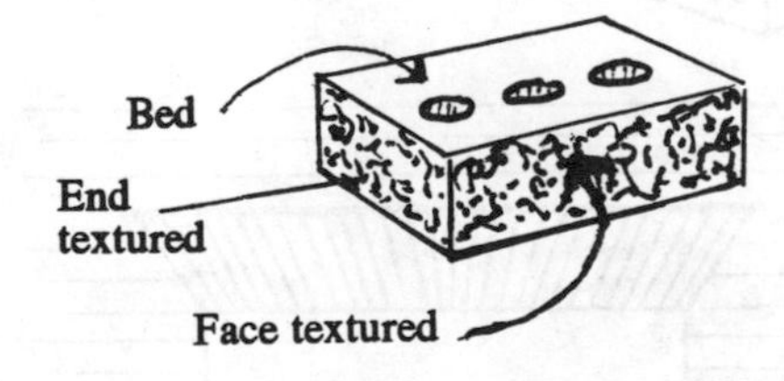

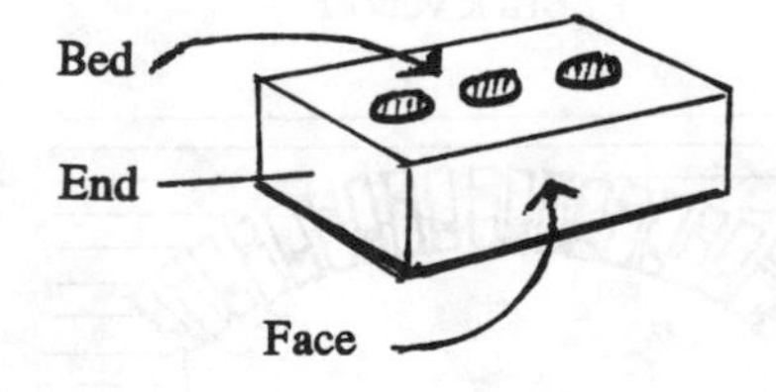

Stone Exteriors

Stone has most of the advantages of brick as well as the disadvantages, but you will not be faced with the problem of the holes in the brick and the texture is uniform. Usually stonework is more costly than brick. If a lot of work with native stone is done in your area, however, this may not be true. Check with your local masons.

Costing

The cost of labor is included in the mason's bid. The cost of the material is provided by the supplier. Don't forget to include the masonry sand, mortar and brick ties.

Management

Have a plan for placement of the material on your lot. If you have very little storage space, you might have to order the delivery of the material in several loads.

Check with your supplier to be sure that existing stocks can provide the brick or stone at your site when you want it and in the quantities you need. If the supplier does not normally stock the brick you have selected, be sure that enough is ordered to complete the entire job and that it is either delivered to your site or held for you. Having to reorder from the manufacturer creates construction delays and may result in a mismatch of color or texture if the second order does not come from the same batch.

If your selection is an oversize brick, the quantities required compared to the standard brick are less on a ratio of about five oversize to seven standard.

Brick ties are galvanized corrugated steel straps about 1″ wide and 4″ long that are used by the mason to tie the brick to the house framing. One end of the tie is nailed to the stud and the other end is buried in the mortar between the brick and stone courses. Most codes require that one of these ties be used for every 4½ square feet of wall area. You will therefore need about 150 ties for every thousand standard brick and about 110 for every thousand oversize brick.

Two or three days after its completion, the brick work should be cleaned with a muriatic acid solution by the masonry crew. Use a ratio of one part muriatic acid to ten parts of water. This work should be a normal part of any masonry work, but verify that your mason has included it in the bid. Muriatic acid is usually supplied in large quantities only, so it is best for your mason, who will be able to use the larger quantities, to provide the acid.

Review Chapter 5, "Foundations," for other management suggestions.

Wood Siding

Solid Wood Siding

Figure 8.3 illustrates the most popular forms of solid-wood siding. Almost any wood can be used for this purpose, including such species as redwood, fir, cypress, pine, cedar, spruce and hemlock.

Lap Siding

Lap siding is normally installed over the sheathing (and poly wrapping, if it has been used) and nailed to the studs. For corner treatment, use either the metal corner pieces or wood corner boards. The thickness of the corner boards should be 5/4″ or 1″ actual. If the corner board is less than 1″ thick, the lap siding may project beyond it. The results will not look good and will be impossible to caulk efficiently.

FIGURE 8.3 Wood Siding

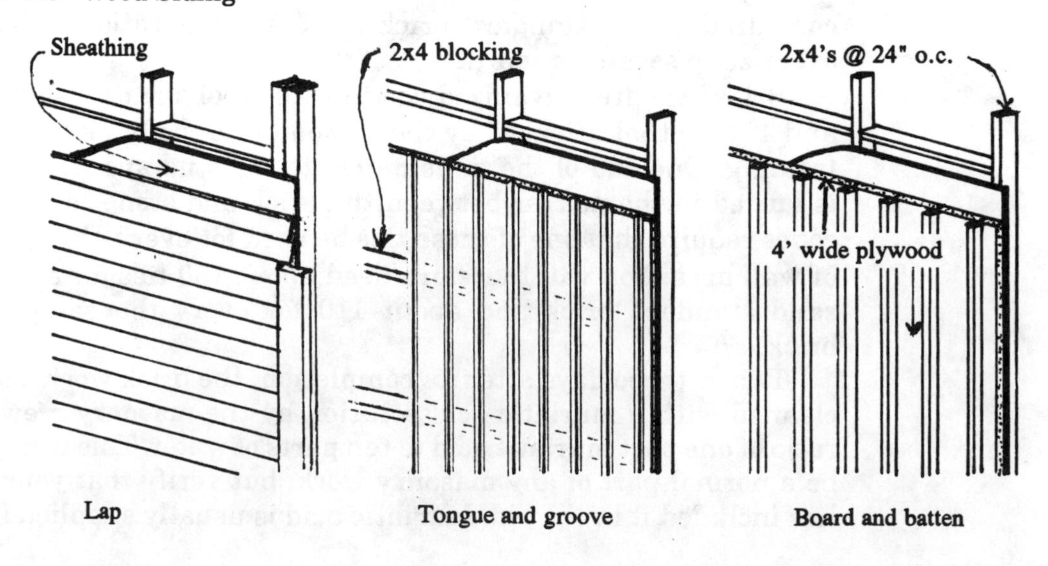

Vertical Siding

To install vertical siding properly, you will have to put blocking between the studs at a vertical interval not greater than 24″. Since most of the pieces of solid wood siding have widths less than the 16″ interval of the studs, this blocking is necessary to provide a good nailing base.

Plywood Siding

Plywood siding can be supplied in many varieties of wood and pattern and at varying costs. Check with your local supplier to see samples. Plywood panels are made in various lengths: 8′, 10′ and 12′. Match the length to the requirements of your house to have as few horizontal joints as feasible. They detract from the overall appearance and can be a source of water and moisture leaks if not properly installed.

There should be a ⅛″ gap in the vertical and horizontal joints between the plywood panels to allow for expansion. Figure 8.4 illustrates methods for forming waterproof joints.

FIGURE 8.4 Horizontal and Vertical Joints in Plywood Siding

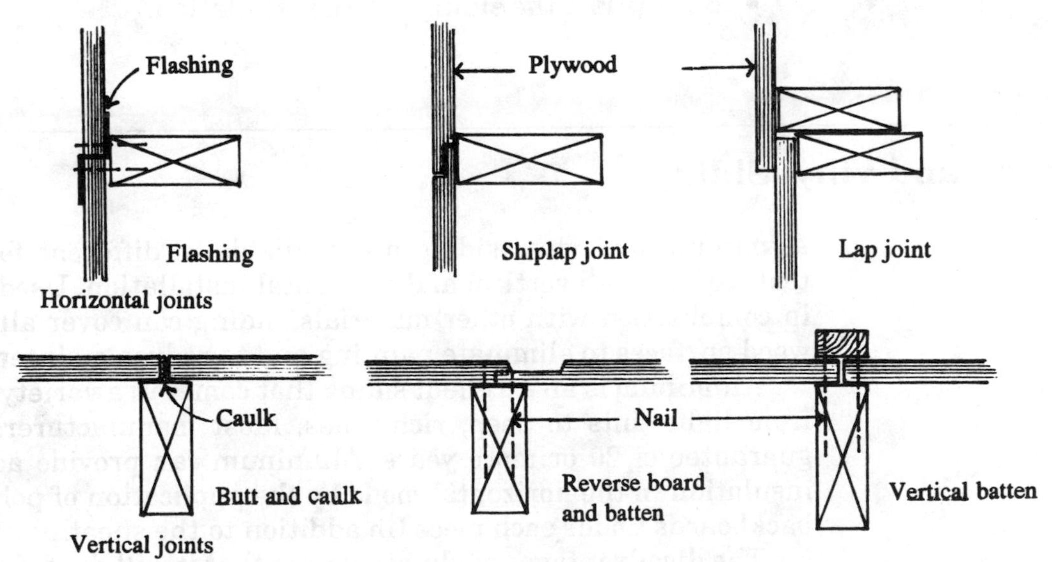

Costing

The price of the materials is provided by the supplier and the labor costs are included in the bid of the carpentry contractor. Expect about 10% waste of material in excess of the actual square footage of the wall area.

Management

- Handle the material carefully on site. It is expensive and may split if treated too roughly. This is particularly true of the relatively thin edges of lap siding.
- Store the material so that it will not warp. Put it on a flat surface and cover from the elements. If your house has a garage, put the siding inside to protect it until installed.
- Do not apply the siding over wet sheathing or studs and do not apply it next to uncured concrete or stucco.
- All-wood siding should be applied with aluminum or hot-dipped galvanized nails.
- As in all phases of the construction, be certain that the siding, nails and related material are on hand before you bring in the carpenters. If the framing crew is to install the siding, find out when the crew chief wants it on the job. Sometimes the siding work can begin before the framing and exterior trim are complete.
- Back prime the siding before installation.

Aluminum and Vinyl Siding

Aluminum and vinyl siding are available in different forms and textures for both vertical and horizontal installation. Used alone or in combination with other materials, siding can cover all exterior wood surfaces to eliminate painting and to reduce maintenance.

Aluminum is an excellent siding that comes in a variety of colors from light tints to deep, rich tones. Most manufacturers offer a guarantee of 20 or more years. Aluminum can provide additional insulation in the horizontal mode by the application of polystyrene backboards under each piece (in addition to the sheathing).

The disadvantages of aluminum are that it will dent if struck by a hard object, and the surface color can be scratched, exposing the bare aluminum.

With vinyl siding, the color selection is limited to the tints. On the other hand, the color appears throughout the entire thickness of the material so that a scratch will do little harm. Vinyl will not dent.

Vinyl is not readily adaptable for covering exposed-wood trim, whereas aluminum is well suited for this purpose. Therefore, if you want vinyl siding and coverage of all-wood trim, use a combination of vinyl for the siding and aluminum for the trim.

Costing

The cost of the material and the labor is normally furnished by the contractor. Many carpentry crews have experience in this line of work. If you choose this option, get the labor cost from the carpentry crew and the material cost from the supplier. Most suppliers are willing to do the material take-off at no cost to you.

Management

- The installation of siding materials requires special skills and tools, and you should be sure that the installers have experience in this type of work. This is particularly true if the job involves on-the-job forming of aluminum to cover exposed wood materials.
- If you are to furnish the material, order so that it is on the job before the installers arrive.
- Be sure that your bid for installation of this type of siding includes the cost of caulking. This task is normally performed by the painting contractor, but with aluminum and vinyl there is no painting involved. Include the caulking task in the specifications when you ask for bids.

Stucco

Stucco is an excellent exterior finish offering long life and requiring very little maintenance. It is a form of concrete and has most of the characteristics of that material. It is made only in white, but can be coated with any paint recommended for application over masonry. You have a wide choice of color selection.

When applied over the sheathing of a wood-frame house, a layer of building paper is placed over the sheathing to protect it from the moisture in the stucco (see Figure 8.5). Next is the application of metal lathing nailed to the studs. The important point in the

FIGURE 8.5 Stucco Wall Construction

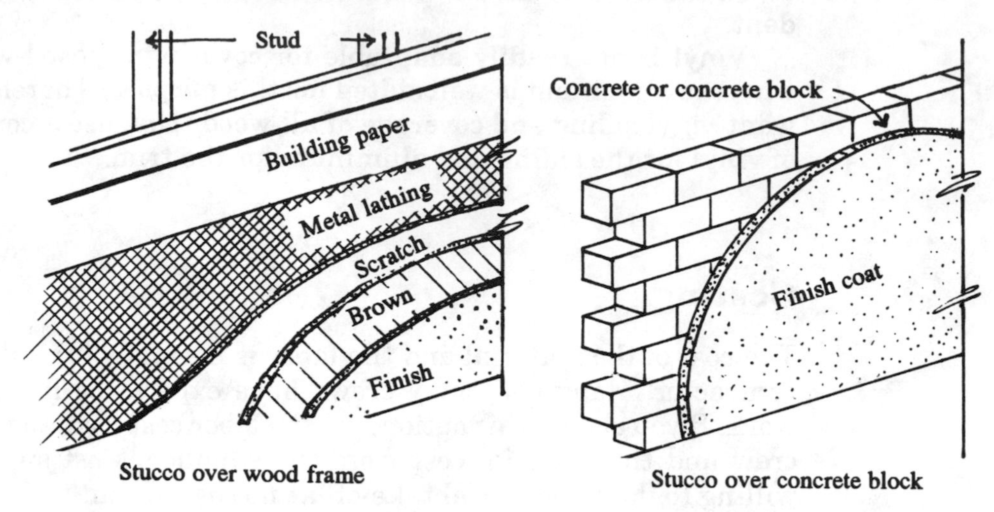

application of the stucco is that the scratch or first coat of material must be pushed through the metal lathing and behind it to form a solid layer between the lathing and the stud. The metal lathing provides a space between the lathing and the stud to permit the *scratch coat* to slip behind and thus anchor itself firmly.

A *brown coat* is applied over the scratch coat after allowing sufficient time for the first coat to dry. And finally, after the brown coat has dried, the *white finish coat* is applied with whatever pattern you have selected—smooth, stippled, swirled, etc.

In applying stucco over masonry, the finish coat is applied directly to the block or concrete. The mortar joints of the block should be flush so as not to provide valleys for the finish coat to fill.

Outside temperature is critical in that during the curing process the stucco coats should not freeze. A safe guide is not to do the job when the temperature is expected to drop below 50° F.

The proper application of stucco requires a special skill. Each step must be done correctly or the process may fail. For this reason, it is best to contract the entire process, the application of the building paper, the lathing and the three coats. It is also preferable that the contractor furnish all of the materials. Where stucco is applied over large uninterrupted areas, control joints should be installed to permit expansion and contraction of the stucco material. Without these control joints, the stucco will crack. As a general guide, control joints should be installed at least every 30′.

Recently, a fiberglass cement board system has become available. It is simpler to install in that a cement board is applied to the outside of the framing over the sheathing. A base coat of stucco is spread on; then the final coat is put on the next day. This system reduces labor costs. It also reduces cracking since this type of material can expand and contract more readily than the standard product.

The finish coat is available in 21 premixed colors (with custom color mixing as well) and four different textures.

Costing

The cost of both labor and material should be provided by the contractor.

Management

- If the contractor supplies the materials and the labor, it is only necessary to specify when the job is ready for the crew to go to work.
- If the contractor prefers that you provide the materials, ask for a list of the quantities needed. You must make sure that these materials are on the job at the proper time and protected from the weather.
- During the stucco process, it is important that no other work is underway, such as hammering inside the house, because it could cause the stucco to fall away from the lathing before the curing process has been completed.

9

Roofing

Shingles

Shingles are the most popular selection for roofing material in homes. However, they should not normally be used on roofs with a pitch of less than $\frac{3}{12}$ or a 3″ vertical rise for each horizontal 12″. At this flat angle, seepage of water, especially in high winds, may occur under the shingles. If installed over a properly applied two-ply underlayment (building paper), shingles can be used on roofs with a pitch of only $\frac{2}{12}$.

Color Selection

If your roof is complex with many dormers and valleys and varying planes, a dark shingle will tend to pull it together.

Check the other houses on the street. If they all use more or less the same color, you may want to select a different one to break the monotony, but do coordinate the color with your house trim.

A light color will reflect heat and thus is more desirable in those areas where air conditioning is the greatest user of energy. In colder climates, since black absorbs heat from the sun, a darker color may be the better choice.

If you live in a mildew-prone area, avoid light-colored shingles since mildew will be most noticeable on them.

Types of Shingles

- *Asphalt* shingles are made in many different colors by many different manufacturers. They vary in weight from 220 pounds per square (100 square feet of roof area) to 340 pounds. The heavier shingle is more expensive, but it has greater texture and longer life. Most roofers also charge a higher rate for applying the heavier shingle.
- *Fiberglass* shingles (single layered) are similar in appearance to asphalt shingles. They are more resistant to fire and may reduce the premium for your homeowner's insurance.
- *Laminated fiberglass* shingles are thicker, heavier and project deeper shadows, which appear more dramatic than the single-layer type. Some designs resemble cedar shakes or slate.
- *Aluminum* shingles are very light and generally have a shake texture. They last indefinitely. Be sure to use aluminum nails for applying them to avoid any electrolytic reaction, which could cause corrosion.
- *Wood* shingles are available in several species, cedar being the most popular. In wood, the term "shingle" indicates the material has been sawn, whereas the term "shake" means the material has been split. The shake is usually thicker and has a much more rustic appearance. The labor to install and the cost of the material can be four to five times that of standard asphalt or fiberglass shingle.
- *Clay tile* shingles are very popular in the sunbelt areas. They are an excellent shingle with a very long life. Tile shingles are very heavy, weighing from 800 to 1,600 pounds per square. If your plans are based on wood or asphalt shingles, have the structural design of the roof checked by an architect or engineer before making the decision to change from standard shingles to clay tile.
- *Slate* is among the finest materials for roofing. It is also one of the most expensive. Again, if this is your choice, be certain the structural design of the framing is based on a slate roof, which can go from 700 to over 3,600 pounds per square.

Ice Buildup

If you are building with a shingled roof in a cold climate, you should consider special construction methods to eliminate ice buildup. Ice dams are caused by heat from the attic space melting the underside of the snow on the roof, as shown in Figure 9.1. The water from the melted snow trickles down the shingles under the snow until it reaches the part of the roof surface over the eaves, which is not

FIGURE 9.1 Ice Buildup

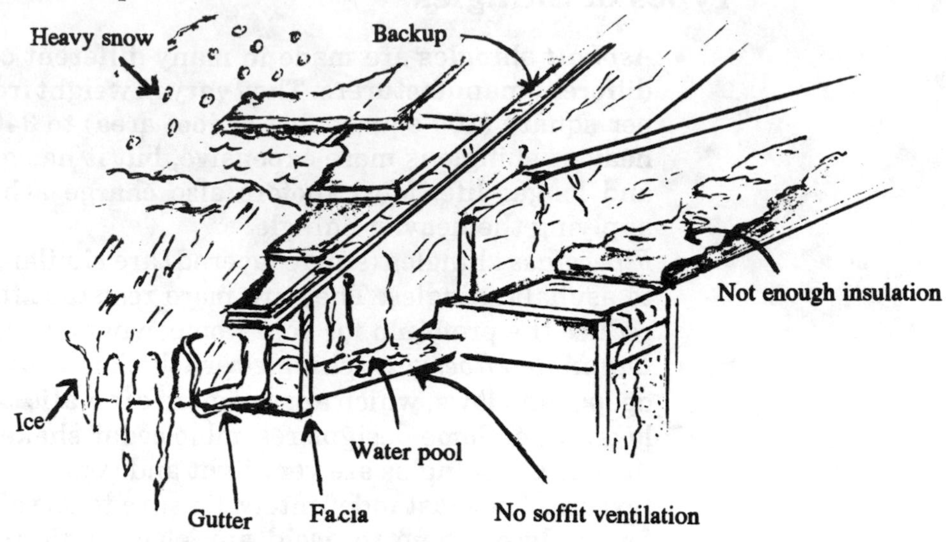

warmed by interior heat. The water freezes and continues to build an ice dam that will move up the roof slope underneath the shingles and leak into the house structure, causing damage to the wood frame, ceiling and wall finishes and other parts of the house interior.

One of the preferred methods of preventing this ice damage is to install rubberized asphalt ice and water shield roll roofing before the shingles are laid. See Figure 9.2 and compare it with Figure 9.1.

A second step is to increase the amount of insulation in the ceiling to reduce the movement of heated air from inside the house into the attic space. The insulation should extend to the top plate around the perimeter of the ceiling. Do not block the movement of outside air through the soffit vents.

A third step, as a last resort, is to install electric heating strips cables along the edge of the roof and into the gutters to prevent the formation of the ice dam.

Metal Roofs

- *Copper* sheeting can be a striking choice particularly on a contemporary home. In time copper turns to a salmon color to brown to black and finally a beautiful green patina. A copper roof can last up to 100 years.
- *Terne* sheeting has been used in America since the 1700s and can last as long as 50 years with proper maintenance. Terne

FIGURE 9.2 Eliminating Ice Buildup

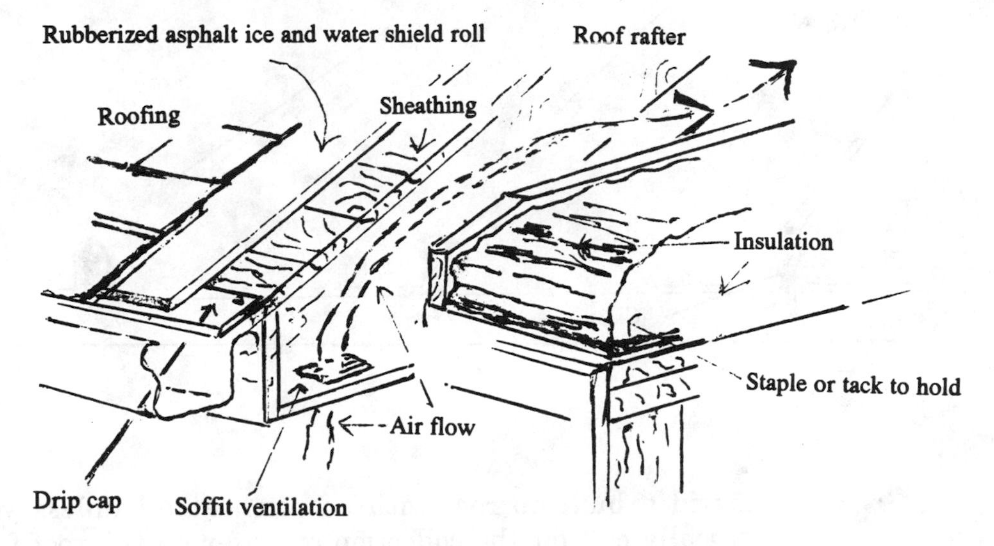

consists of a lead-tin coating over a copper-bearing steel base. Typically, Terne must be brush painted with a linseed-oil primer when installed and repainted every eight to ten years. It takes and holds paint extremely well.

- *Terne-coated stainless (TCS)* sheets consist of stainless steel coated with Terne alloy. It does not have to be painted, but it can be if its normal gray color is not desired. TCS lasts at least 30 years.

Sheet-metal roofing is usually installed in a standing seam configuration (see Figure 9.3).

Metal roof sheeting should have no penetration through the sheeting. This is accomplished by installing hold down cleats that are formed into the standing seams.

Built-Up Roofs

Figure 9.4 illustrates flat roof construction consisting of several alternating layers of fiberglass felt with a coat of tar in between and a finish coat of gravel on top.

Another newer system uses a single synthetic rubberlike membrane instead of the fiberglass felt. It is less affected by the sun's rays and readily expands and contracts with the roof as the temperature changes. Most roofers are familiar with this product.

FIGURE 9.3 Standing Seam

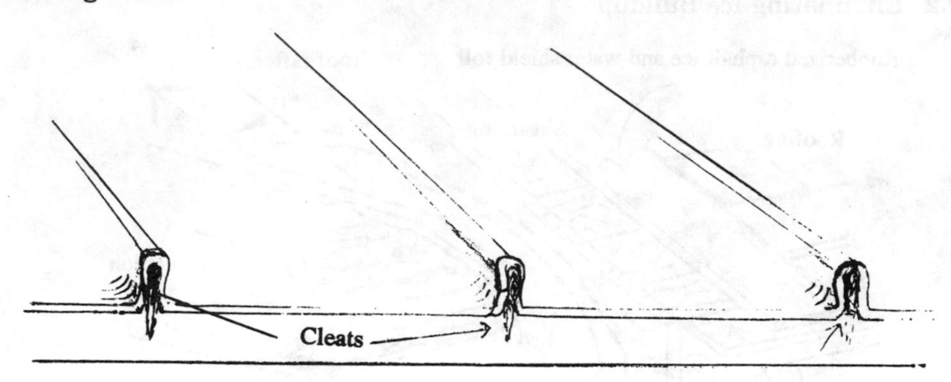

Avoid a built-up roof with no pitch at all unless your plans specifically call for the collection of water on the roof for heating and/or cooling. Under these circumstances the roof must be specifically designed to hold the water. The built-up roof with no pitch will collect and hold water and tend to develop leaks.

Flashing

Flashing is the application of material at critical points, such as the juncture of the shingles with the chimney or the juncture of the siding of a dormer with the main roof, to prevent water and snow

FIGURE 9.4 Built-up Roof Construction

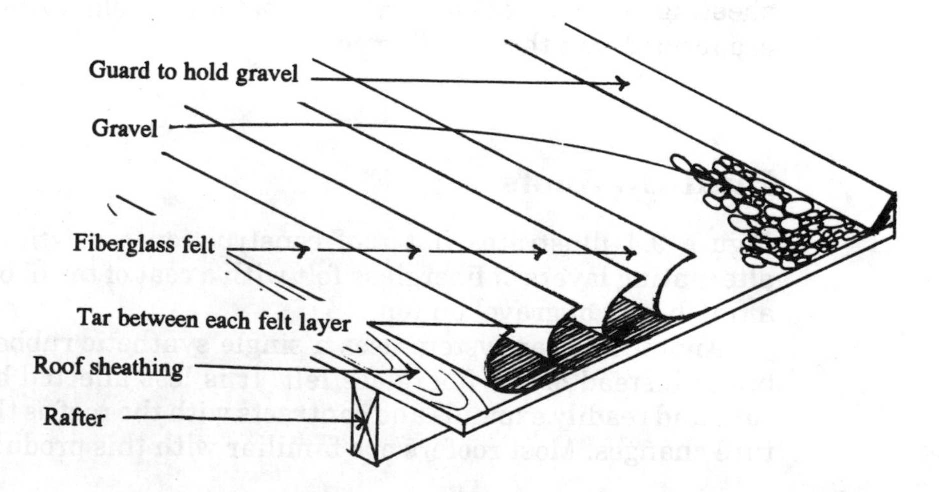

from leaking through the roof system. Figure 9.5 illustrates some flashing applications.

An excellent choice for the flashing material is copper. It seems to last forever and requires almost no maintenance, but is expensive. A vinyl material in various colors is now on the market. It is an excellent choice and costs less than copper. Aluminum can also be used, but it has the drawback of a shiny appearance unless painted, and it does not hold paint well. Galvanized and tin-plated steel can be used, but they also must be painted.

Gutters and Downspouts

Gutters and downspouts can be useful in preventing drainage problems. They are a vital necessity, in certain cases, to carry water away from the house. They can be very troublesome in areas with many trees, since they will require cleaning four or five times a year.

Cleaning gutters can be facilitated if you specify removable end caps. Then you can pop off the cap and flush the gutter with a garden hose. Also require that the gutters be mounted so that the back of the gutter is offset from the facia board. Without this air space, the paint will fail and the facia board will rot.

Your choices of guttering materials are copper, aluminum with baked on paint, galvanized steel, which must be painted, and vinyl. Considering both cost and long life with low maintenance, the best choices are vinyl and aluminum.

Underground Roof Drains

If, in your area, it will be necessary to clean out your gutters several times a year to remove leaves, acorns, seeds and other yard debris, underground drains may be your best choice. See Figure 9.6.

The drain consists of a ditch about 1′ deep from grade level and 1½′ wide centered directly under the roof edge where the rain and snow will run off.

Place a continuous 4″ slotted plastic pipe in the center of the bottom of the ditch. Fill the remainder of the ditch with gravel. The individual stones of gravel should be large enough so that they will not move into the slots of the pipe and reduce the flow of the drain water.

With this system, the water falls from the roof edge down to the gravel in the ditch, then into the plastic pipe where it is carried away to a storm drain, stream, pond or other collection source.

FIGURE 9.5 Roof Flashing Details

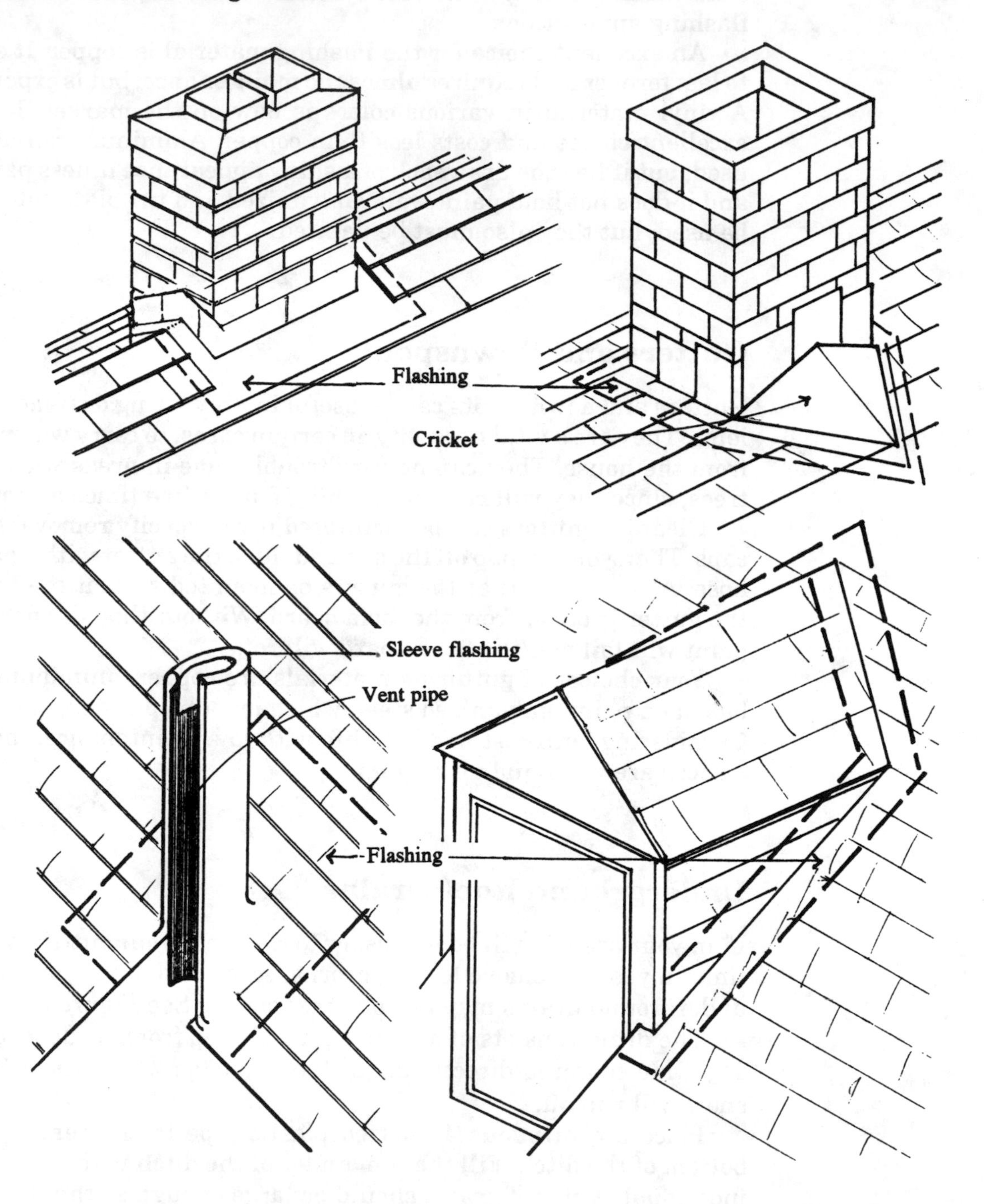

This system will function even when covered with leaves, twigs and other yard debris.

Costing

The cost of roofing labor is provided by the contractor. Either of you can furnish the shingles or other roofing material. Just be sure that this is made clear in your bidding.

Flashing material and labor to install it are usually provided by the roofer, but you must specify the type of material. The plumber will furnish and install flashing around plumbing vent pipes that penetrate the roof.

The roofer usually supplies the labor and material to install gutters and downspouts. Make certain that your plans indicate the actual length of downspouts required, including any changes to foundation height.

If you have built-up roofing, ask the roofing contractors to include both labor and materials in their bids.

FIGURE 9.6 Underground Roof Drain

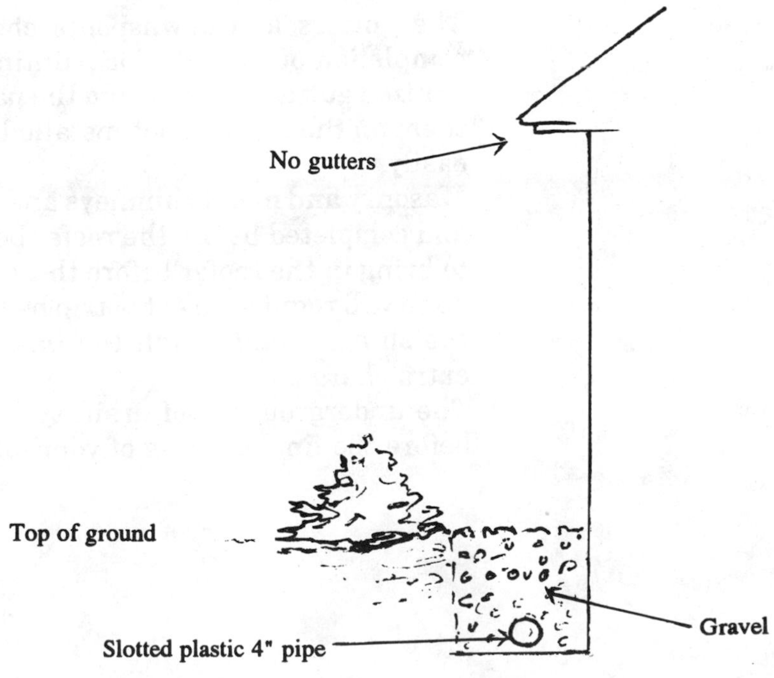

No gutters
Top of ground
Gravel
Slotted plastic 4" pipe

Labor for installing an underground roof drain system can be provided by the contractor who put in the footings. The supplier can give you the cost of the pipe and gravel.

Management

- The shingles should be installed very soon after the application of the building paper on the roof by the framing crew. In rough, windy weather this paper will not last long. In addition, you should push to have the house weatherproofed as soon as possible.
- If you are to furnish the shingles, make sure that they are on the job at the proper time. Arrange with your roofer for a location for the supplier to place the shingles on site. Like most material, shingles are heavy and the contractor doesn't want to have to move them any farther than necessary.
- Avoid having shingles placed on the lot way ahead of the time when they will be applied, particularly in hot weather. The mastic that seals the shingles after installation will sometimes, during open storage, soften and leak past its cover strip and stick to adjacent shingles. This leakage can also happen if the supplier has old stock, so make sure that your shingles are fairly new.
- The gutters and downspouts should not be installed until completion of the exterior painting. If you have selected galvanized gutter material and the painter is to paint them, have them on the site but not installed so that the job can be done easily.
- Masonry and metal chimneys and skylights should be in place and completed before the roofer begins. It is not good practice to bring in the roofer before these items are finished, since to do so will require an extra trip by the roofing crew to complete the shingles and install the flashing later. It will justify an extra charge.
- The underground roof drain system should be installed just before the final grading of your lot.

10

Masonry and Concrete

This chapter discusses those items of masonry and concrete that have not been included elsewhere. Some of the items mentioned are overlooked in the costing of a house.

Fireplaces

For centuries, the fireplace functioned as the principal means of heating a house and in later years as an auxiliary heating system. Unfortunately, it has not done the job very well. Depending on the use of room air for combustion, the fireplace can actually waste more heat than it generates. In modern homes, the room air used for fireplace combustion has already been heated by the primary heating system, but most of it is lost up the chimney. Therefore, rather than assist the primary system in its heating function, the fireplace increases the work load of the primary system.

To be energy efficient and to provide a realistic addition to the primary heating system, the fireplace must have:

- A *combustion chamber,* which is supplied with fresh air from outdoors
- A *heat chamber,* which is warmed by the combustion but is isolated from the outside air and the combustion chamber

Air heated by the fireplace is blown from the heat chamber to the room and then back into the heat chamber for reheating. Air to support the combustion of the fuel is drawn from out of doors through special ducting into the combustion chamber (sealed from the room by glass doors) around the fuel, then out through the damper and up the chimney. With this design, heat from the room is not lost up the chimney and the fireplace becomes an efficient heating system on its own or an efficient auxiliary system to the primary system (see Figure 10.1).

Types of Fireplaces

The all-masonry fireplace With a full masonry fire box and chimney, this fireplace can be designed and built for energy-efficient operation, as indicated above. Its construction requires a skill that may not be widespread since it is based on a new technique. The labor costs will be high, but maintenance low and the beauty of the brick work can enhance the overall appearance of your room. If you choose this option, be certain that you have a good design and that your mason has had experience in building this type of fireplace.

The masonry/steel-box fireplace This type is similar to the all-masonry fireplace except that the fire box is prefabricated steel. The remainder of the system is all masonry. This type is less costly and retains most of the other advantages of the all-masonry fireplace in

FIGURE 10.1 Energy-Efficient Fireplace Design

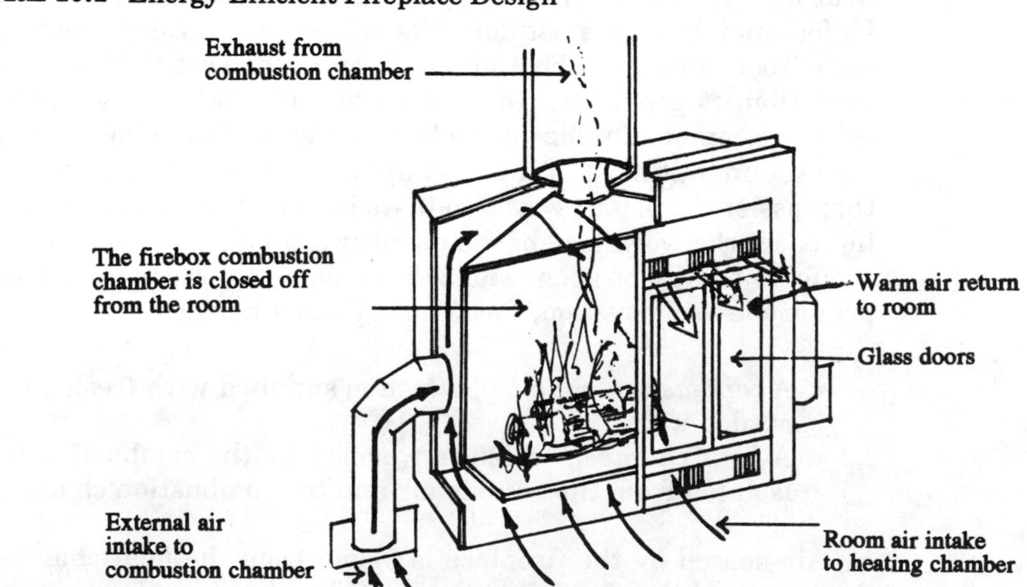

regard to architecture and beauty. Vents into and out of the combustion chamber can emerge from the front or side of the box.

The all-steel fireplace This type is 90% prefabricated, the only remaining requirement being the assembly and installation of the various parts. The system includes the fire box, the parts of the steel chimney in lengths of 2′, 3′ or 4′, the chimney cap and related equipment. This type of fireplace has been on the market for some time, and most framing crews are experienced in its installation. The fireplace functions efficiently and is the least costly of the three systems. It may present one difficult problem, however: what to use to cover that portion of the metal chimney that protrudes through the roof.

If your house has exterior wood, aluminum or vinyl siding or stucco, even in part, one acceptable method to conceal the exposed metal chimney is the use of 2×4 framing (size to complement the architecture of the house) covered with the same siding used on the rest of the house.

The top of the frame must be covered by a metal cap to prevent rain from getting inside the structure. Because this cap is not normally furnished by the supplier of the fireplace, it must be job made. Contact your roofing or heating contractor (both usually have sheet metal tools) and have them make it for you.

If your house is brick or stone veneer, the weight of this material is such that it is not practical to use it to hide the wood-framed metal chimney. You may use the factory made artificial brick covers, but they do not always look very attractive. Your best choice is to select the all-masonry or the masonry/steel-box fireplace, which has the brick chimney built from top to bottom.

The room-side finish on steel fire boxes can be brick, tile, plaster, stucco or almost any material that does not create a fire hazard.

Most building codes require a minimum height of 3′ for that part of the chimney above a flat roof. If your roof has any pitch, this exposed chimney height could be several more feet depending upon the amount of the pitch. For most codes, the rule is that the chimney must be 2′ higher than the point of the roof that is 10′ away (measured horizontally) from the chimney. In Figure 10.2, note that the height of the exposed chimney not including the cap is almost 10′.

Electrical Requirements for the Fireplace

All of the fireplaces described use electric blowers to circulate the heated air from the heat chamber to the room and then back into the heat chamber for reheating. Be sure that your electrician is

FIGURE 10.2 Chimney Height Requirements

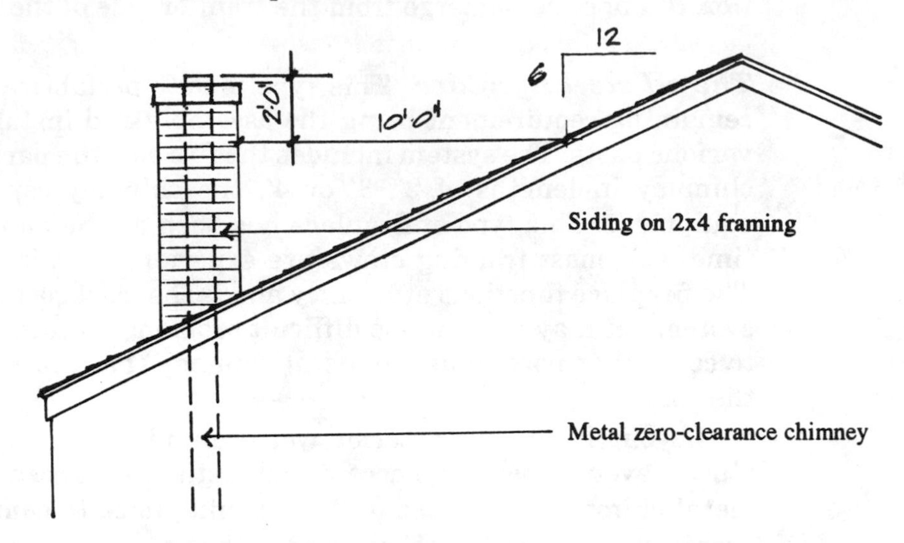

aware of the need to provide the hookup. In addition, select a design that will allow easy access to the blower for periodic maintenance. Most blowers do produce some noise even when set on "low," so choose one that can be turned off completely while the fireplace is operating.

Masonry Hearths

If your fireplace is built with a hearth of some form of masonry or tile at floor level, the weight of this material can normally be carried by the floor joists with some modest reinforcement. On the other hand, if your fireplace has a large raised hearth, the weight of the masonry can be considerable. It must be supported from the ground up beginning with a concrete pad at least 12" thick (usually part of the total fireplace footing) and then brought up to the subfloor level with concrete block. Failure to provide this type of support may cause separation of the heavy, large hearth from the fire box after the house is built.

Fireplace Mantels

If your fireplace is to have a heavy timber mantel, check the local building code to see what limitations you must follow for its installation. Usually, as long as the timber mantel is placed 12" or more above the fireplace opening, the amount of the projection does not matter.

Most milled wood mantels are built to conform to these rules and present no installation problem.

Gas-Fired Fireplaces

Efficient prefabricated gas-fired fireplaces are available today in many styles; they are easy to install and much easier to take care of than the wood burners. Many types can be installed directly against a plaster or sheetrock covered wall or on top of an existing wood floor. Look for the "zero clearance" type. Most of these fireplaces need a very simple "chimney" consisting of a short 4″ metal pipe installed horizontally through the exterior wall behind the fireplace into the outside air.

If public gas service is available in your area, you can hook up to it for the fuel supply. If not, butane gas can be supplied from an exterior tank.

Gas-fired stoves with similar characteristics are also available.

Miscellaneous Masonry

Walks, patios and other exterior slab work should not be built until the final grade has been determined and preferably finished. Among the more popular materials used for this work are:

- *Brick* laid either in mortar on a concrete slab or laid without mortar on a level bed of crushed rock or sand. The mortarless system requires less skill and is therefore one task you can undertake yourself. The preferred brick for these jobs are called pavers and are supplied in thicknesses of 1½″ to 2″. If you would rather use the brick matching that on the house, don't forget the problems of color and texture previously discussed in Chapter 8, "Exterior Finishes." Order solids to eliminate the problem of holes in the brick.
- *Stone or flagstone* can be laid in mortar or directly on a sand bed that has been tamped and leveled. If the work does not require mortar and concrete, consider doing it yourself.
- *Concrete* is an excellent material for patios and walks. Choose any of four selections: standard concrete, standard concrete with color added to the mix, exposed aggregate concrete and concrete stamped with a pattern.

FIGURE 10.3 Exposed Aggregate Concrete Patio

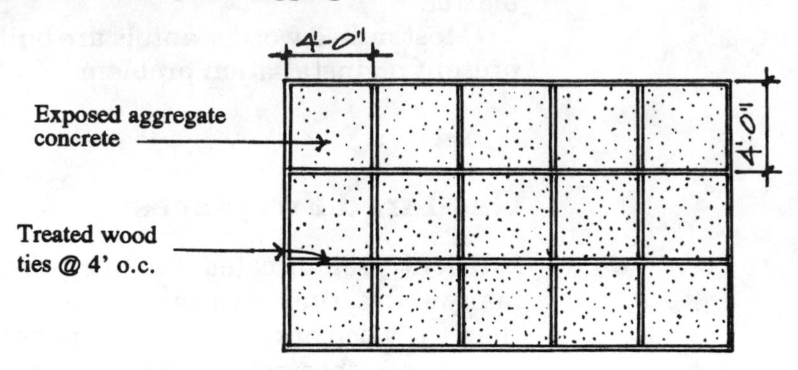

The aggregate of exposed aggregate concrete consists mainly of brown smooth pebbles. Shortly after the concrete is poured and floated, a retardant is applied on the surface. This liquid holds back the setting of the top film of the concrete. At the appropriate time (following the directions of the retardant manufacturer), the top layer is washed with water, thus removing the concrete binder and exposing the top of the brown pebbles. This process gives the concret an attractive textured look.

Concrete stamping is the technique of using special tools to imprint designs into the top of the concrete to give the effect of tile, brick, cobble stone and so forth.

Another interesting method of laying concrete is the use of 1″ or 2″ boards (redwood, cedar or salt treated pine) as expansion joints (see Figure 10.3). The wood inserts also serve as forms for the concrete and facilitate screeding (leveling).

FIGURE 10.4 Concrete-Slab Porch Floors

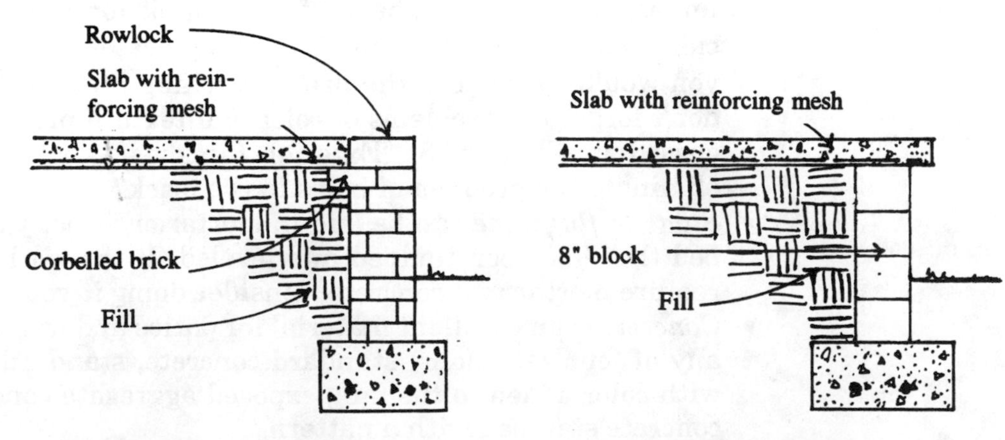

Concrete Porch Floors

Concrete porch floors should be supported by the masonry wall that contains the fill and not by the fill (see Figure 10.4). This design will prevent any dropping or cracking of the concrete floor due to further compaction or settling of the fill. Make sure that in each case the metal reinforcement in the concrete reaches over the corbeled brick or over the block wall.

Porch floors may also be constructed of the different concrete options discussed earlier, or of earth stone, ceramic tile, or almost any other form of masonry.

Retaining Walls

Retaining walls can be made of most forms of masonry, landscape timber or railroad ties (see Figure 10.5).

The masonry must be laid on suitable footings. These walls must have provisions to drain water through them by the use of weep holes or by a plastic footing drain pipe to prevent pressure build up behind. The types of walls illustrated are to be used only for jobs where the height of the wall does not exceed 4′.

FIGURE 10.5 Retaining Walls

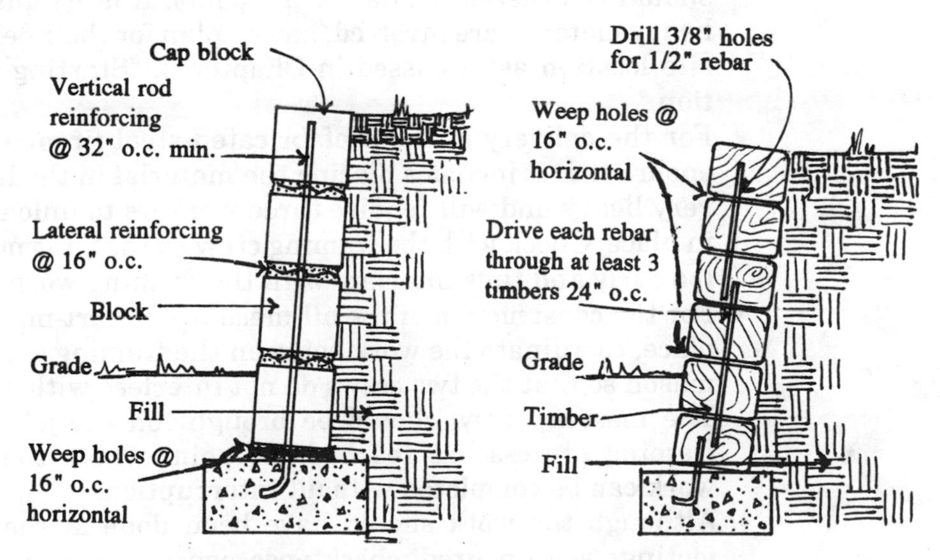

Costing

- In computing the cost of the all-masonry fireplace, have the mason give you the material requirements and the labor cost. Be sure that the estimate includes the brick, block, mortar, firebrick, flue liners, damper and brick ties.
- In computing the cost of the combination masonry/steel fireplace, have the mason give you the material requirements and the labor cost. The suppliers of the fire box and masonry will give you the cost of their materials.
- For the all-steel fireplace, the cost of the parts is provided by the supplier, the labor cost by the installer. Don't forget to include the labor and materials for whatever treatment of the exposed chimney you have chosen.
- Costs for other masonry, such as sidewalks and patios, are derived from the labor charged by the mason and the material cost provided by the supplier.
- For wood retaining walls, the supplier will give you the cost of the materials and the contractor the labor.

Management

- Have the materials on hand at the proper site location before the work is to be done (except the concrete, of course, which should be ordered the day of the pour). If large quantities of heavy material are involved, have a plan for their delivery and site location as discussed in Chapter 3, "Starting Construction."
- For the delivery of the prefabricated-steel fireplace system, ensure that it includes placing the material in the house. It is very heavy and will require three workers to unload and put in place. Check with the framing crew so that it is not ordered too early and thus interfere with the framing work.
- For the construction of the all-masonry or part-masonry fireplace, coordinate the work between the framing crew and the mason so that the two crews do not interfere with each other. The masonry crew should be brought on the job when the framing process has reached the point where the masonry work can be completed without interruption.
- Although the work should have been done at the time the footings were poured, check once again to ensure that adequate concrete footings are in place for fireplaces and masonry steps.

11

Air Infiltration

One of the most important elements in producing an energy-efficient home is to reduce air infiltration to the minimum. If special sealing measures are not taken during construction, large quantities of air will move back and forth through cracks in the walls, floors and ceilings of the occupied house. Studies produced by HUD, the National Association of Home Builders, the National Research Council of Canada and the National Applied Science Development Center indicate that the average loss of house energy through infiltration is about 33% with some test homes experiencing a loss as high as 60%.

During the heating season, warm indoor air holds more moisture than cold outdoor air. This creates vapor pressure, which constantly forces warm water vapor out through exterior walls, floors and ceiling.

Fortunately, sealing a home against air infiltration is a simple process *if done during construction,* and it is relatively inexpensive, particularly considering the energy saved.

Most of the sealing is accomplished after the framing is complete and the electrical, plumbing and heating rough-ins have been finished and inspected. Hold off on the insulation until the sealing has been completed with the exceptions discussed in the remainder of this chapter.

The job should be performed by an experienced sealing contractor who will apply the proper sealant at each place. In most cases, the material used is an expanding foam sealant, which is excellent for filling permanently large openings such as the gap between the frame of the window and the stud framing of the house and in caulking small cracks in selected areas most likely to leak.

Where To Seal

Figures 11.1 and 11.2 illustrate those areas to be sealed. The numbers 1 through 8 show in detail just where the sealant is applied.

1. **Seal between the sill plate and the foundation wall. This sealing should be accomplished during the framing by the carpentry crew installing a special batt of insulation made for this purpose. Install this batt even if the inside of the foundation is ventilated crawl space, since it will reduce the effect of the wind in the winter months.**

FIGURE 11.1 Sealing for Air Infiltration

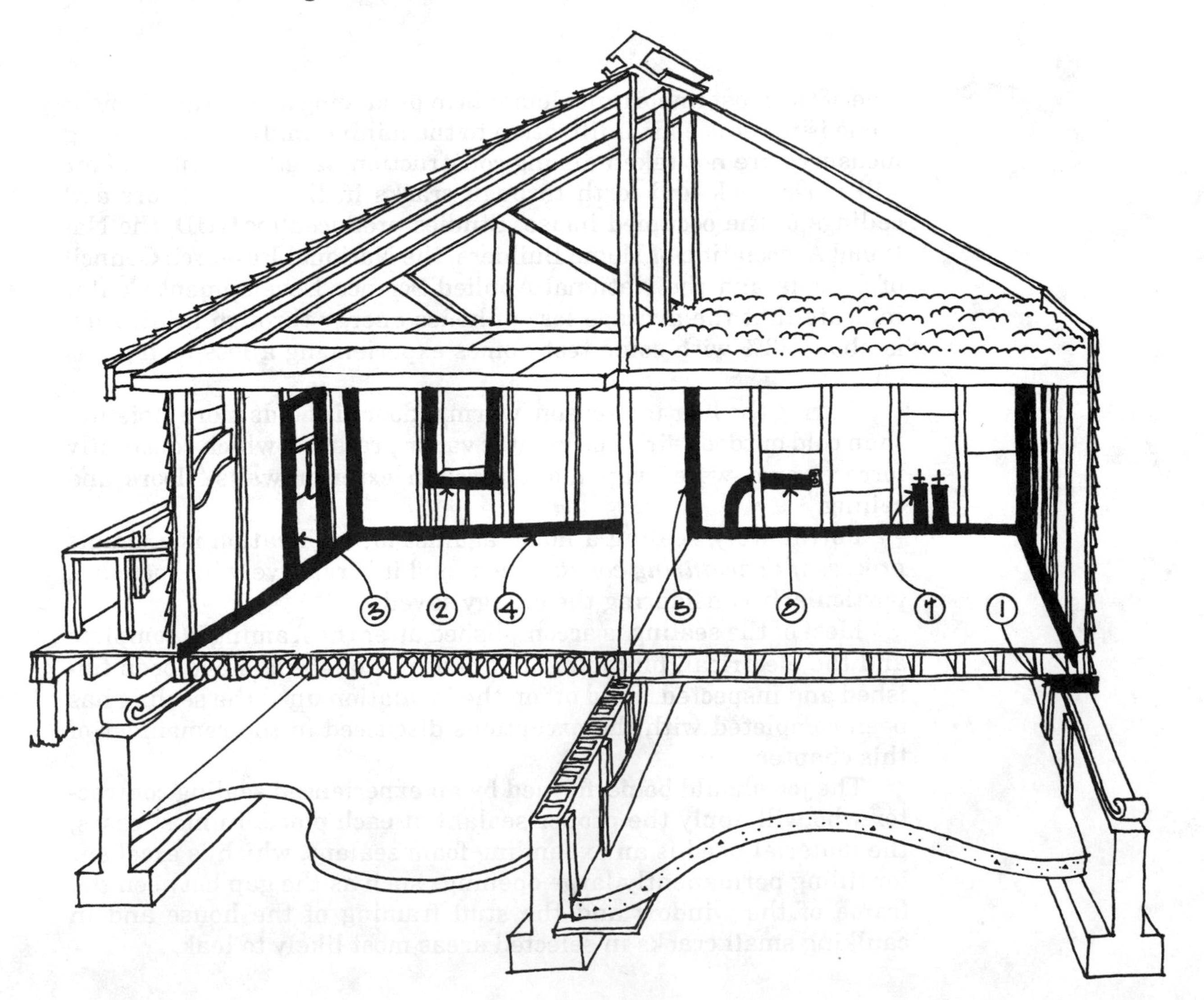

FIGURE 11.2 Sealing for Air Infiltration: Details

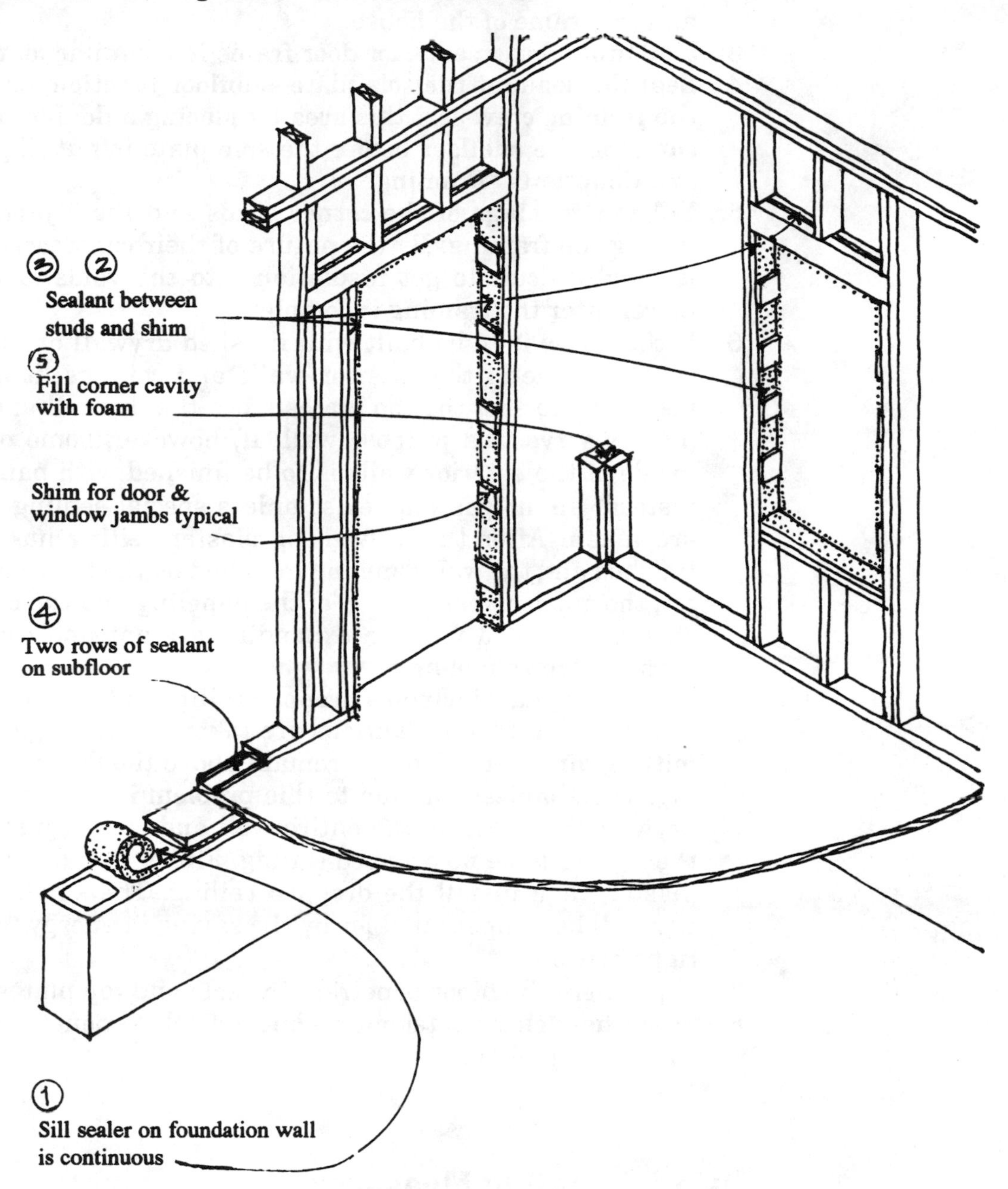

2. Seal around the window between the frame of the window and the frame of the house.
3. Seal around each exterior door frame in a similar manner.
4. Seal the seam at the sole plate–subfloor junction, or have the framing crew seal this area by placing a double row of caulk on the subfloor before the sole plate is put in place. See Chapter 6, "Framing."
5. Fill the openings of the corner studs and the T-junctions during the framing. By the nature of their construction, it is very difficult to get insulation into the voids of these pieces after the framing is complete.
6. If the house is to be built with finished drywall or plaster on the inside of the exterior walls and ceilings, it is not necessary to seal the top plates since the drywall and the plaster serve this purpose well. If, however, some of the inside of the exterior walls is to be finished with paneling instead, an air gap may exist unless special sealing steps are taken. After the drywall or plaster ceiling has been finished and the wall paneling installed by the trim carpenter, the gap between the top of the paneling and the ceiling should be sealed with an expanding sealant and covered with the crown molding.

   If your plans include a dropped ceiling in the kitchen or elsewhere, unless precautions are taken a serious gap permitting air infiltration will remain above the dropped ceiling. The simplest solution to this problem is to install the drywall or plaster on the entire wall and ceiling just as if there were to be no dropped ceiling. Finish this drywall or plaster, then install the dropped ceiling. The small extra cost will be compensated for by the saving in energy from a tight house.
7. Seal where the pipes penetrate the sole and top plates.
8. Seal where electric, telephone and TV cables penetrate the sole and top plates.

Additional Sealing Measures

The most effective anti-air-infiltration measure is to wrap the entire exterior wall, over the sheathing before the siding is applied, with material such as DuPont's Tyvek; there are several other similar materials. These wrappings are paper thin, very strong and will prevent the passage of air through the exterior walls without trapping moisture within the walls. The cost for this installation is very low compared with the fuel savings for heating and cooling. Make certain that the framing crew repairs all damage to the sheathing before wrapping.

Seal around plumbing drains under the bath tubs by blocking large openings with pieces of construction sheathing and sealing the remaining cracks with expanding foam.

The object of this entire sealing process is to enclose the whole living area of the house to minimize the passage of air, heated and cooled, through the exterior walls, ceilings and floors and thus decrease the cost of heating and cooling the house.

Costing

The sealing contractor should provide the cost of both labor and materials in the bid.

Management

- Schedule the contractor at the proper time.
- Following the guidelines discussed in this chapter, check the work to ensure that none of the areas to be sealed have been overlooked.

12

Insulation

The basic principle of insulation is the capture of air space and the use of this air space as a barrier to the movement of heat. Consequently, the less dense the material (i.e., the greater the air space in the material), the greater its insulating value. For example, in

FIGURE 12.1 Thicknesses of Materials for Same Insulation

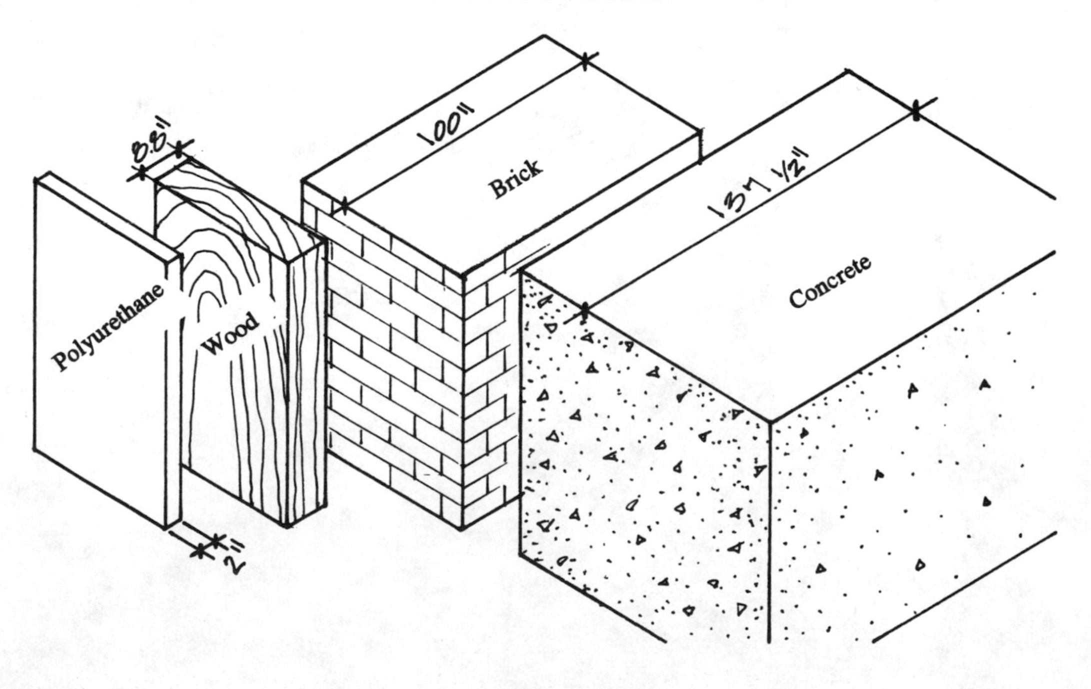

Figure 12.1 the 2″ thick polyurethane, has the same insulating capacity as the 137½″ thick concrete.

In house design, the living area (the area heated and/or air conditioned) should be completely enclosed by the proper amount of insulation for the local environment. This process consists of the application of insulation in the exterior walls, ceilings and under the floor for both crawl space and concrete slab construction.

R–Value of Insulating Material

The ability of material to resist the passage of heat is expressed in terms of R. For example, a fiberglass batt 6″ thick has an R–value of 19. The same material in batts 12″ thick has an R–value of 38. The greater the R, the greater the material's resistance to the passage of heat.

R–Values of Various Insulation Materials

	Batts		Loose and Blown Fill		
R-value	*Fiber-glass*	*Rock wool*	*Fiber-glass*	*Rock wool*	*Cellulose fiber*
R–11	3″–4″	3″–3½″	5″	4″	3″
R–19	6″–6½″	5–6″	8½″	6½″	5½″
R–22	7″–7½″	6–7″	10″	7½″	6″
R–30	9″–10″	8″–9½″	13″	10″	8½″
R–38	12″–13″	10″–12″	17″	13″	10½″

U–Value of Insulating Material

On occasion, you will find the thermal quality of an insulating material expressed in terms of U. The U–value is the reciprocal of R and can be determined by dividing the R–value into one. For example, the fiberglass batt with the R–value of 19 has a U–value of $1/19$ or .053. The lower the U–value the greater the thermal resistance of the material. U–values are used primarily by engineers, architects and heating contractors in determining the heating-cooling system needs for an entire structure.

Vapor Barrier

In addition to insulation, the living area should also be sealed with appropriate material applied to the inside of the studs as well as the ceiling and floor joists to prevent the movement of moisture from the living area into the insulation.

The vapor barrier serves two purposes. First, insulation will lose some of its thermal properties if it becomes damp or wet and, second, if the moisture is retained inside the living space, the occupants will be more comfortable with less heat. This second reason does not apply to the use of air conditioning, particularly in humid climates, because one of the principal functions of an air-conditioning system is to remove moisture from the air.

There are three methods for applying a vapor barrier to exterior walls.

1. The first is to install insulating batts with vapor-barrier backing such as treated kraft paper or aluminum foil with the vapor-barrier material on the living area side. If you use loose blown-in insulating material in the ceiling (usually less costly than batts), establish a vapor barrier as described in the following two paragraphs.
2. Second, use friction fit or another type of insulation without an integrated vapor barrier and install aluminum foil backed drywall. This method provides a good vapor barrier and is simple to put up. In those rooms where gypsum paneling is not used as a wall finish, use either the system described above or below.
3. Third, use friction fit or another type of insulation without an integrated vapor barrier and glue, staple or nail polyethylene sheet material to the interior of the studs and ceiling joists. This is probably the most effective barrier, but there is some question among experts whether or not the increased efficiency of this system compared with the aluminum backed drywall is worth the extra step of installation and the extra cost. In addition, if this system is used, drywall, wood or other paneling cannot be applied with adhesive, the best method for attaching these materials. (See Chapter 17, "Interior Walls and Finishes.")

With regard to floors, the usual procedure for installing an effective vapor barrier in floors over crawl space is to place 15-pound building paper between the subfloor and the finished floor. In addition, apply batt insulation under the subfloor between the floor joists with an integrated vapor barrier faced up against the sub-floor.

If your house is built on a crawl-space foundation, make sure that a vapor barrier is placed over the soil after all construction operations using the crawl space have been completed. This measure prevents deterioration of floor framing because of high moisture levels.

Installing Insulation

A properly insulated house may use several types of insulation to isolate the heated living area from the outside atmosphere. Figure 12.2 shows examples of where the various types can be installed.

1. Shows the use of 3½″ batts or 6″ batts with kraft paper vapor barrier in stud walls (The vapor barrier faces the heated area.)
2. Illustrates the installation of friction fit batts, unfaced, i.e., without vapor barrier (A vapor barrier of polyethylene film not less than 3 mils thick has been applied over the insulation on the heated side of the wall.)
3. Ceiling using batts
4. Blown-in loose ceiling insulation
5. Insulation between foundation wall and sill plate (This serves the purpose of stopping air infiltration more than it insulates, which is what it should do.)
6. Underfloor insulation (Vapor barrier must be placed on the side next to the heated space, or "up.")
7. Insulating basement areas (This insulation, installed between the studs or furring, should be covered with drywall or paneling as with any interior wall. The studs or furring (should be salt treated .25CCA) lumber are fastened to the block or concrete walls with case hardened nails or nails driven by an explosive charge. It is also a good idea to paint the exterior masonry walls, before the studs are put up, with two coats of a masonry waterproof paint.)
8. Installing rigid insulation (about 2″ thick) under the concrete floor (The entire floor need not be insulated; only the 2′ around the perimeter with a vertical piece of the rigid insulation against the wall the same thickness as the concrete floor. This insulation is, of course, applied before the concrete floor is poured. Also, review Chapter 5, "Foundations.")
9. Stapling reflective foil (shiny aluminum foil with reinforcing layers or fibers is preferred) on the bottom of the roof joists to reduce the amount of heat entering the attic space (If your roof has a metal ridge vent, make sure that the vent is not covered by the foil so that air flow from the soffits to the ridge vent moves freely.)

FIGURE 12.2 Installing Insulation

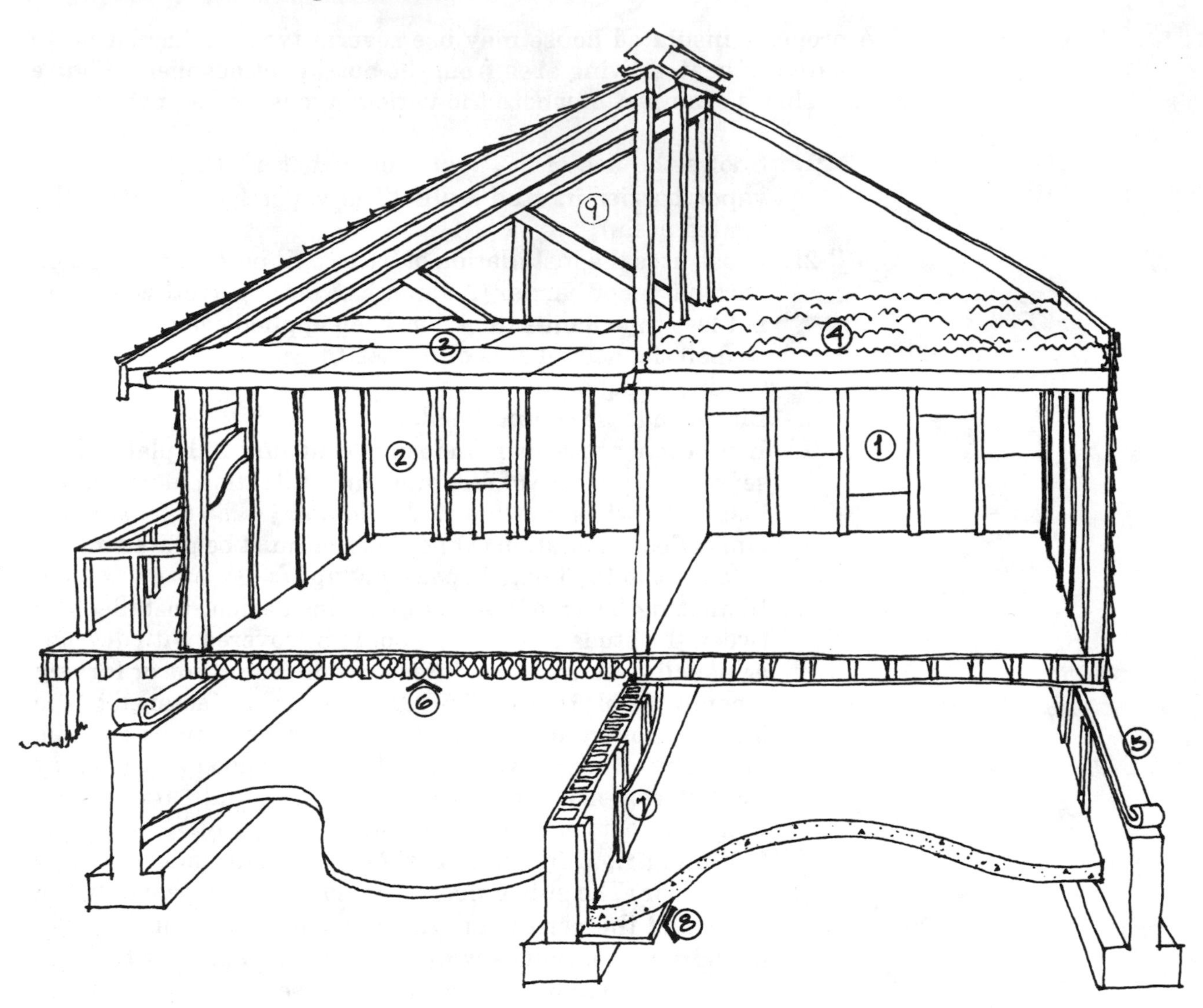

Insulation of Cathedral Ceilings

In the construction of a cathedral ceiling, the ceiling joists associated with the normal 8′ high ceiling are eliminated so that the insulation cannot be placed between the joists over the drywall. There are two methods used in solving this problem.

Figure 12.3 illustrates the construction of a cathedral ceiling in which the dry wall is nailed to the underside of the roof rafters. False beams can be added under the sheetrock for decorative purposes. In this type of ceiling, the insulation must consist of batt material, which is nailed to the bottom of the roof rafters *before the installation of the drywall.* Note the air space between the top of the

FIGURE 12.3 Cathedral Ceiling with Drywall or Plaster Finish

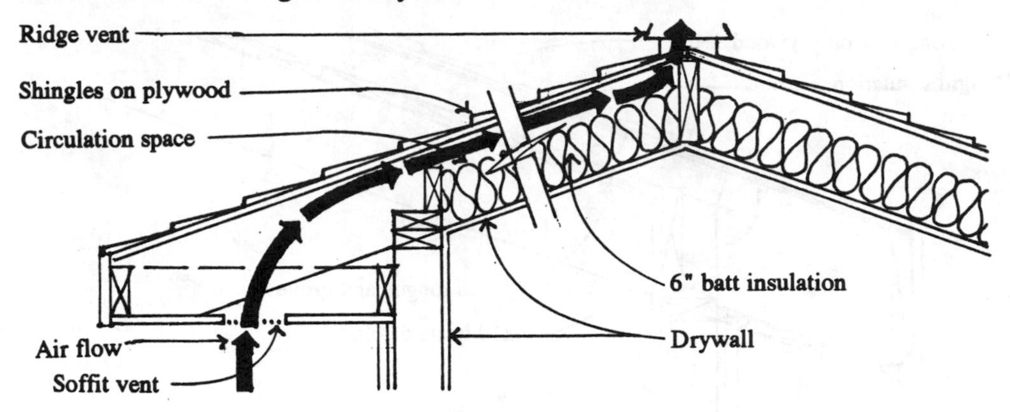

insulation and the underneath side of the roof sheathing. This space is very important to ensure the proper ventilation of the area. To allow for this air space, the roof rafters must be sized larger than the required depth of the insulation. For example, if the roof insulation is to be 6″ batts (just about the minimum for most of the country), the roof rafters must be at least 2×8 to allow a 1½″ air space. If the required insulation is 8″ batts, then the roof rafters must be 2×10s, and so forth.

Figure 12.4 illustrates the type of cathedral ceiling associated with the post and beam framing found in many contemporary designs. In this construction, the roof sheathing consists of tongue and grooved lumber, at least 2″ thick. This sheathing is nailed directly on top of the roof rafters (usually at least 3″ thick material) with the underside of the sheathing and the rafters forming the finished ceiling. In this case, the insulation will consist of rigid urethane nailed to the outer side of the sheathing with building paper, hard board for nailing and shingles installed on top. The ventilation problem does not exist because there is no confined air space.

R–Values of Rigid Urethane Board for Roof Insulation

1½″ thick board = R–7
2″ thick board = R–12
3″ thick board = R–21
4″ thick board = R–29

Higher R–Values can be attained by combining various thicknesses of boards.

FIGURE 12.4 Cathedral Ceiling: Exposed Beam and Plank

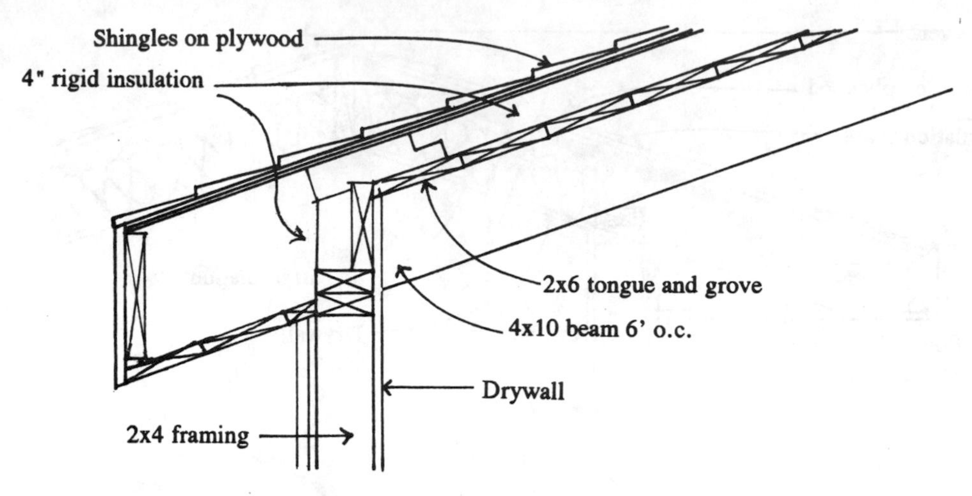

Figure 12.5 indicates the various R–Values for ceilings, walls and floors. Those building in the higher-numbered (colder) areas and Alaska and Canada should examine the construction techniques in Appendix F, "The Superinsulated House," for ways to obtain greater energy savings.

One of the most neglected areas for potential heat loss through both air infiltration and lack of insulation is the pulldown stairs to

FIGURE 12.5 Recommended R–Values for Areas in the United States

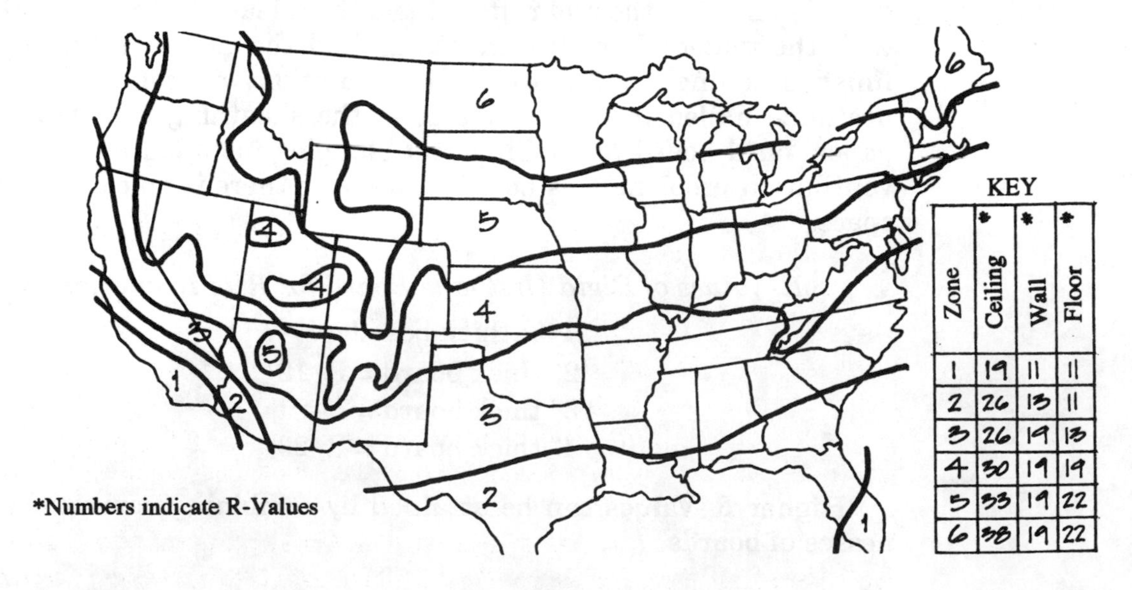

FIGURE 12.6 Insulation for Pulldown Stairs

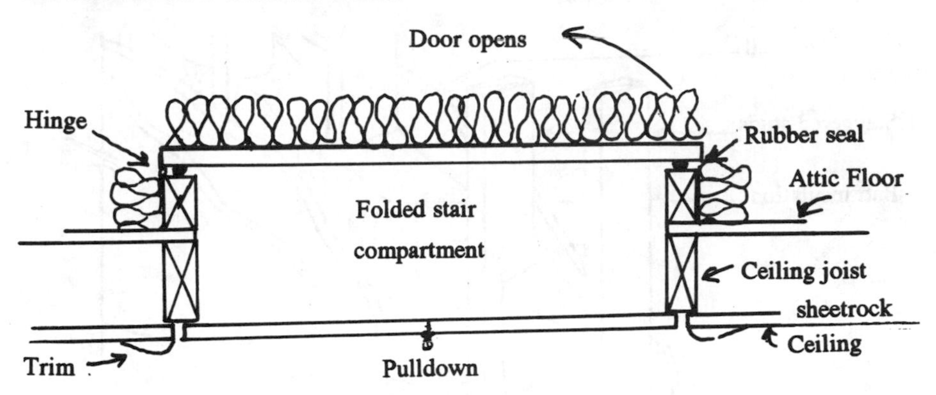

the attic space. To install these items properly requires special measures (see Figure 12.6).

The frame installed on top of the subfloor must be high enough to accept the pulldown stair in its folded position with about an inch of clearance. The door should be hinged on one side of the frame and equipped with screen door hooks so that when closed, the door can be snugly pulled against the rubber seal to stop air infiltration.

Insulation to match the other parts of the ceiling should be installed on top of the door and around the sides of the frame. Seal the seam at the juncture of the frame and the subfloor with caulking compound.

A similar installation should be built around the openings of ceiling-mounted attic fans used for house ventilation. The door is opened when the fan is in use.

Sound Insulation for Walls

If your plans include special sound deadening in certain interior walls, such as between the bathroom and living room, two sound paths must be blocked. One is through the air space in the wall and the other is through the studs themselves.

Figure 12.7 illustrates construction techniques that will do both. Note that the wall is built with staggered studs of 2×4s on a 2×6 plate. This technique eliminates the touching of the drywall on both sides of the wall by any stud and thus reduces the passage of sound through the material itself. The air space voids are filled with 3½″ batts. Note that one side of the wall is finished with two layers of ½″ drywall. All drywall should be the fire-rated type, which provides

FIGURE 12.7 Sound Insulation for Walls

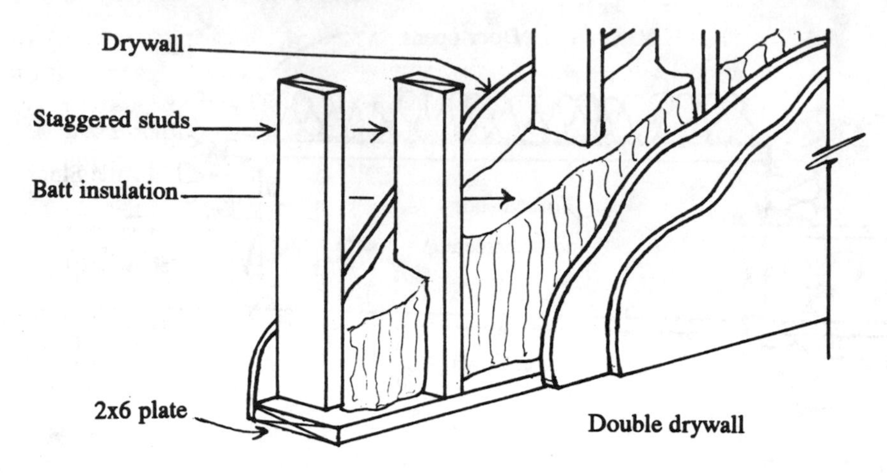

a better sound-deadening barrier. This combination gives an excellent Sound Transmission Class (STC) of 53.

If your house is to use a metal such as copper for water supply pipes, reduce the noise of flowing water by packing felt around the pipe where it goes through the framing or masonry walls as shown in Figure 12.8.

FIGURE 12.8 Sound Insulation for Metal Pipes

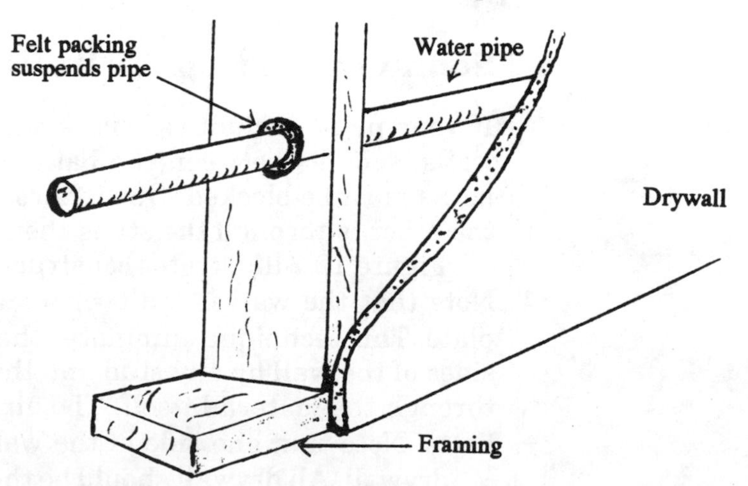

Sound Insulation for Floors

Passage of sound through floors is not usually a problem for single-family ranchers without basements; it is more typically a requirement for apartments and multifloor houses. Should your plans call for a noise barrier, Figure 12.9 illustrates a method of accomplishing it.

Starting at the top is a layer of carpet followed by its pad, a particle board underlayment of 5⁄8″, the subfloor of 1⁄2″ plywood, 31⁄2″ batts of insulation without vapor barrier and finally a layer of 1⁄2″ drywall installed by nailing to a special resilient channel. The drywall should be fire rated since this type provides better sound insulation. The resilient channel is a narrow steel form nailed to the ceiling joists and parallel to them. The drywall is nailed or screwed to the channel on one side only. Your drywall contractor should be familiar with this process. Installed in this manner, the resilient channel will tend to dampen the sound and deter its transmission between floors. This system will give a Sound Transmission Class of 53, which is excellent.

Insulating Block Walls

If your plans require block walls or block walls with brick or stucco exteriors that need to be insulated, the methods shown in Figure 12.10 can be used.

In method 1, use lightweight concrete block and fill the cavities in the block with loose insulating fill.

FIGURE 12.9 Sound Insulation for Floors

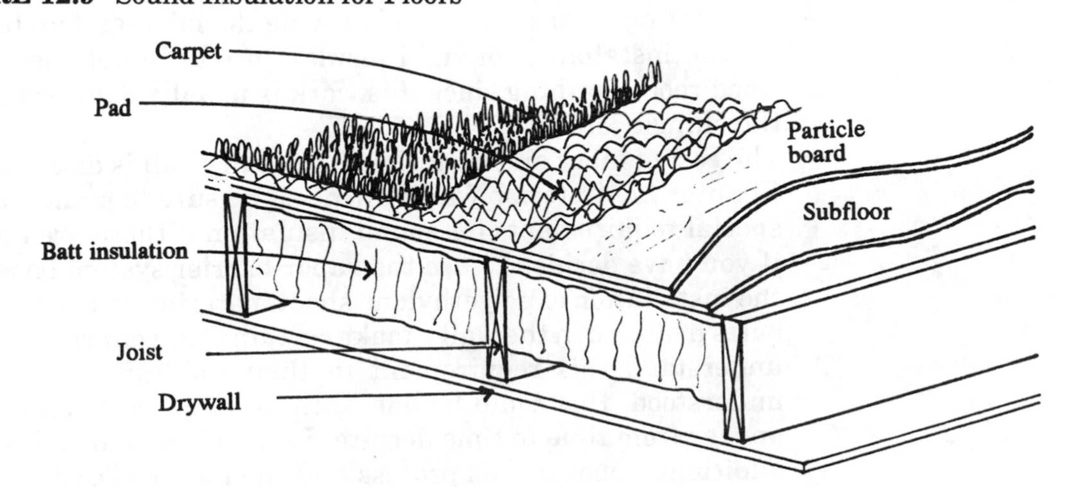

FIGURE 12.10 Insulating Block Walls

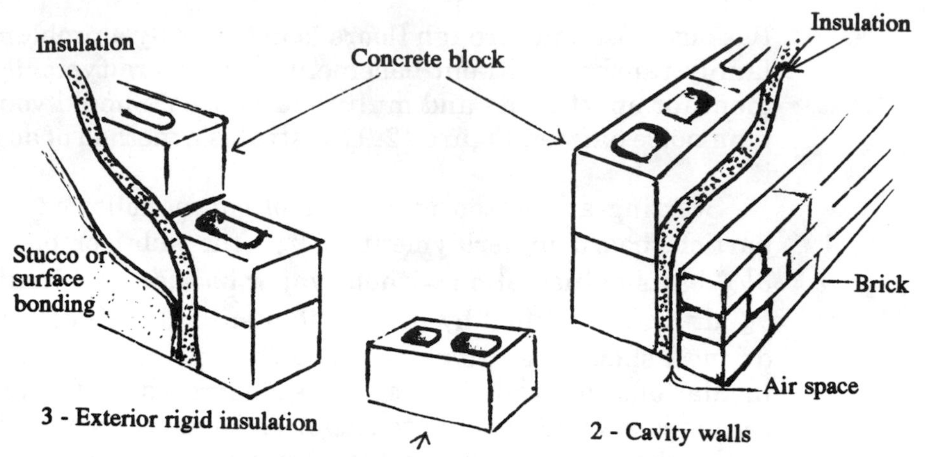

In method 2, construct the block wall as described in method 1, install rigid insulation, then an air space and brick veneer.

Method 3 is designed to accommodate stucco or surface bonding. Construct the concrete block wall with rigid insulation as in method 1, and then apply the stucco or surface bonding directly to the rigid insulation.

Costing

- The cost of both labor and materials for the insulation work is provided by the contractor.
- The roofing contractor should provide the labor and materials for the installation of rigid insulation on the outside of the wood roof sheathing since this work is usually done as part of the shingle job.
- The cost of labor and materials for the drywall is discussed in Chapter 17, "Interior Wall Finishes." Be sure to point out the special requirements for sound insulation if there are any.
- If you have decided to use the vapor-barrier system based on the installation of polyethylene sheet material over insulation batts installed in the walls, make certain that your contractors understand this requirement in their bidding. If it is not understood, this could be one of those extra cost items that appear from time to time despite the best laid plans. There is additional labor in this process and the material costs more.

Management

- Make sure that your insulating contractor is an experienced one. The skills and know-how required to do a good insulation job may not be readily apparent.
- Schedule the contractor to move in just after the sealing has been completed. And once again, *be sure that the rough-in inspections of the framing, electrical, plumbing and heating have been completed and approved.*
- At this stage of the job, the insulating contractor will only install the wall insulation. Until the ceiling drywall is applied, the insulation for standard ceilings should not be put up.
- If you have cathedral ceilings of the type shown in Figure 12.3, the insulation must be installed at this stage. Insulation for the type of cathedral ceiling shown in Figure 12.4 should be installed by the roofer at the point where the job is ready for the shingle work. The roof insulation is part of the overall roof system; therefore it is best to have one contractor do the whole job. In case trouble should develop later, you have to deal with only one contractor.
- The insulation under the floor in the crawl space should not be applied until most all of the crawl space work, such as final inspections for electrical, plumbing and heating, have been completed. This scheduling prevents damage to the insulation by other contractors.
- Since they tend to be trouble spots, you should personally check the following (see Figure 12.11):

    1. Look for gaps, breaks and voids in insulated areas. In particular, make sure that the insulation for the cathedral ceiling has no gaps. After the ceiling drywall has been installed this insulation is completely covered and very difficult to check.
    2. Check to see that the soffit to attic ventilation has not been blocked.
    3. After the underfloor insulation has been installed, check that the headers (band board) around the exterior walls have been covered.
    4. If your plans call for cantilevered overhangs, such as the second floor jutting out beyond the first or bay windows jutting out from exterior walls, be certain that insulation has not been omitted from the underside of these areas.

- Management of the drywall installation is discussed in Chapter 17.

FIGURE 12.11 Attic Ventilation, Sill Plate, Cantilever Insulation

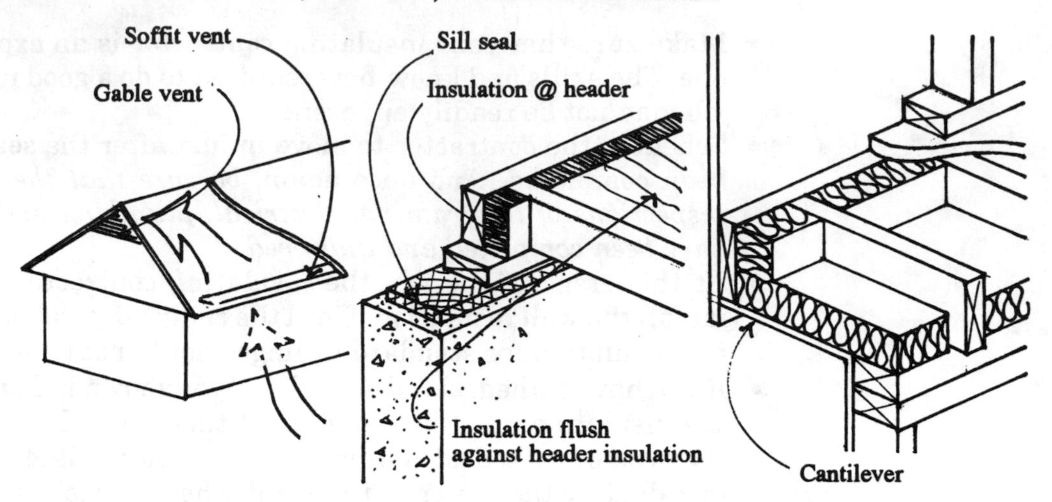

13

Solar Heating

Solar energy is free. Therefore, any time it can be used for efficient heating and cooling, do so. In the present state of the art, solar energy has its limitations; it does not produce economic and efficient heating in all its forms and in all geographic areas of the country. In most situations, a backup heating system is required.

It is important to understand that heat moves in the following three ways:

1. *Conduction* is the way heat moves through a solid, such as metal or wood or masonry. If you place a metal spoon in a cup of hot coffee, the heat will move up the handle. This is conduction.
2. *Convection* is the way heat moves through the air, water and other liquids. Warm air rises because it is lighter than cold air. This causes heat to accumulate at the ceilings and in the second floor of a house.
3. *Radiation* is the movement of heat as a wave similar to light. For example, an electric resistance heater mounted in the bathroom wall will emit heat radiation that travels toward the cooler areas of the room.

Passive Solar Heat

A passive solar heating system uses the natural processes of conduction, convection and radiation to move collected heat. Its functioning is based primarily on the proper structural design of the house. With the exception of the occasional use of low-powered electric fans to assist in air distribution, it does not use mechanical equipment. It is the most practical use of solar energy in the house. To function efficiently, however, the house must be designed specifically to use this type of heat in your area. Any passive solar heating system should have the following components:

- A *collector,* usually consisting of large glass or plastic window areas facing generally south or at least within 30° of south. During the heating season, the collector should not be shaded by other structures or trees from about 9 A.M. to 3 P.M.
- An *absorber* element, usually consisting of a masonry wall or floor or a water tank. The absorber, a storage element, receives the heat that has passed through the collector and holds it for later distribution.
- A *distribution system,* which takes the heat from the storage element and circulates it throughout the living area. A strictly passive system would use all three methods of heat transfer. Some passive solar heating systems require the addition of electric fans to help circulate the warm air.
- A *control* or *regulation device* to prevent the loss of heat from inside the house to the outside during no sunshine periods. Most heat regulators consist of movable insulating curtains. A control system may also include electronic sensing devices to signal a fan to turn on or to open or close vents and dampers that restrict the flow of heat.

Figure 13.1 illustrates the simplest form of passive solar heat—direct gain. Sunlight enters the house through the large window

FIGURE 13.1 Direct Gain

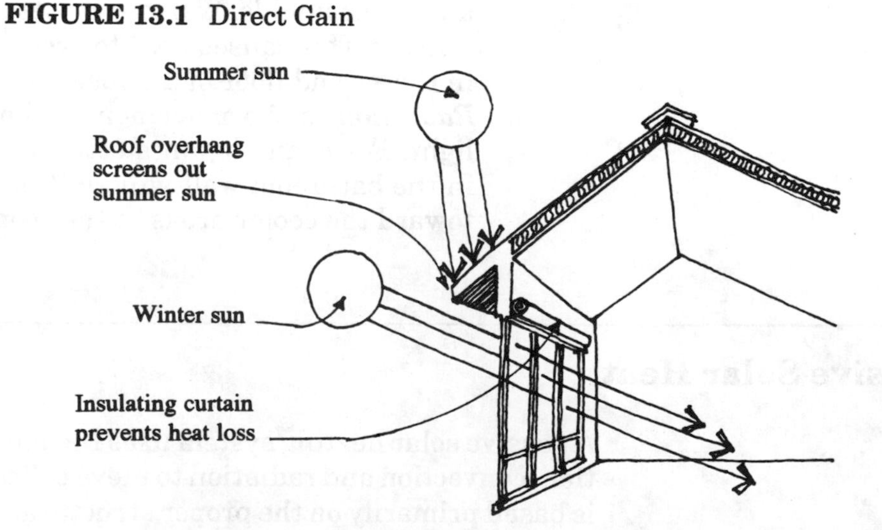

area, the collector, striking the walls and floor where it is absorbed and stored. To be effective, the walls and floors should be built of masonry or water tanks to provide the amount of storage capacity needed. They should also be painted or covered with material of a dark color to improve the absorption process. At night as the room cools the heat stored in the walls and floors radiates into the room. Note also the insulation curtain, which should be closed at night and during non-sunshine days to prevent the loss of heat back through the collector. This is one of the drawbacks of the direct gain passive solar heating system. It gives the owner two poor choices: either lose the heat through the collector or draw the curtain and lose the pleasure of daylight.

The Trombe Wall

Figure 13.2 illustrates the Trombe wall (named after its French inventor). This is a form of indirect gain. Its principal difference from the direct gain is that a solar heat storage wall is placed directly behind the glass collector with a gap of only a few inches. This design provides a much more efficient system than direct gain.

Heat passes through the collector to the storage wall, which should be about 16″ thick, with a dark coating on its surface. Over a period of time the heat will migrate through the storage wall (conduction) into the living space when the temperature in the living room falls below that of the storage wall. In addition, the passage of heat into the living space can be attained quickly by designing the Trombe wall with vents, as shown. As the air between the collector

FIGURE 13.2 The Trombe Wall

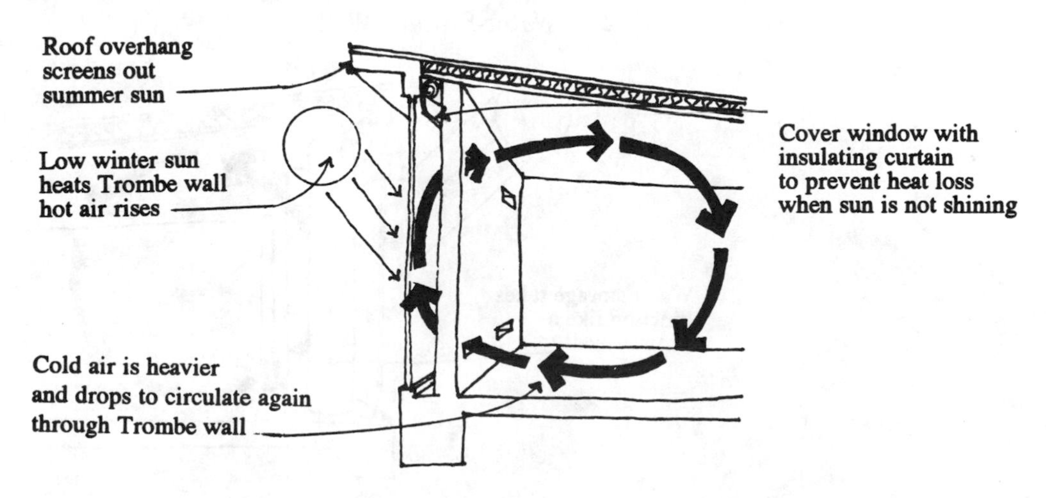

and the storage wall heats up, it will begin to rise and flow through the ceiling vents heating the living space. This flow creates a draw that pulls in cooler air from the living space through the floor vents and thus sets up a pattern of natural air movement—a convective loop. Note the insulation curtain to control heat loss during no-sunshine periods. In addition, the Trombe wall vents are equipped with back-draft dampers to prevent reverse flow at night and thus cool the room air.

One obvious disadvantage of the Trombe wall is the loss of view through the windows.

The Water-Storage Wall

In a variation of the Trombe wall, the masonry storage wall is replaced by a water storage wall (see Figure 13.3). The water can be held in specially designed metal or plastic tanks or in 55-gallon drums adapted for this use. The water wall will absorb more heat than the masonry wall and thus provide a more efficient system. Note that the insulation curtain is retained as one of the important parts of the design.

The Solar Greenhouse

Figure 13.4 illustrates the solar greenhouse concept. Not only can it be used in new construction, but it is also easy to apply in updating older homes. If your preference for architecture is along the more

FIGURE 13.3 The Water-Storage Wall

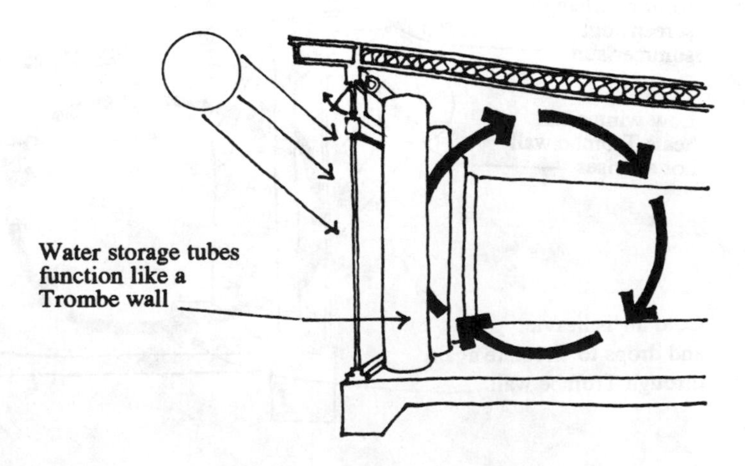

FIGURE 13.4 The Solar Greenhouse

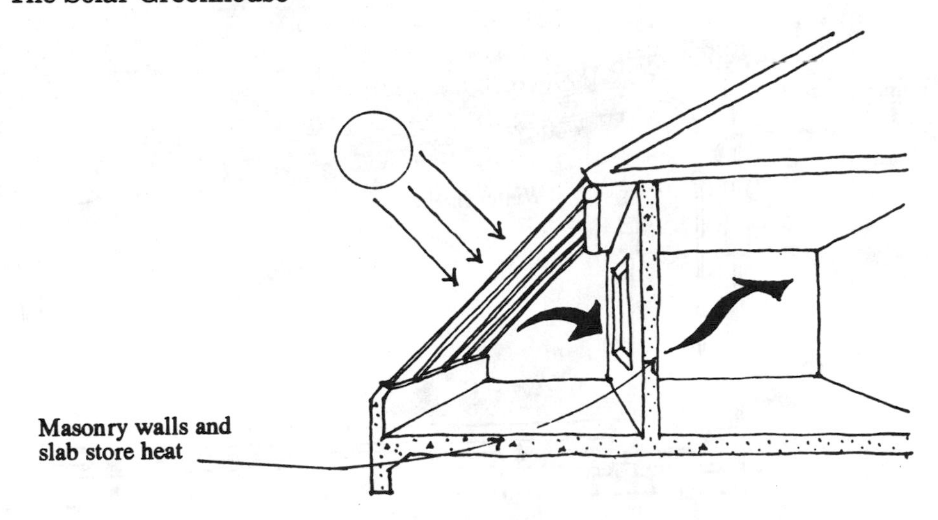

traditional lines, the solar greenhouse can be readily used without major modifications to the overall architectural style.

Solar heat is collected in the greenhouse, then absorbed by the masonry and the water-filled tanks, which can be positioned at random. Distribution of the heat from storage can be made in a fashion similar to the Trombe wall and the direct gain concept. Heat can also be moved from the greenhouse by opening the doors or windows between it and the living area. Low-powered fans can be used to improve the efficiency of circulation. Again, a movable insulating curtain is required. One additional bonus of the solar greenhouse design is that, in fact, it does provide a greenhouse for growing plants or additional space for many other purposes.

The Envelope House

Figure 13.5 illustrates the envelope house concept of passive solar heat in which the measures previously discussed are put together into a package providing a living space around which heated air is circulated by passive means (in most cases supplemented by low-power circulating fans) to produce a house of high energy efficiency.

Both the top and bottom of the attic space are insulated so that the heat is retained by the air as it circulates through this space. Since the crawl space is used as a heat-carrying chamber, it must be insulated. Crawl-space vents are eliminated.

The theory of the envelope house is that the sun heats up the air in the greenhouse and the air rises to the attic space. In the north

FIGURE 13.5 The Envelope House

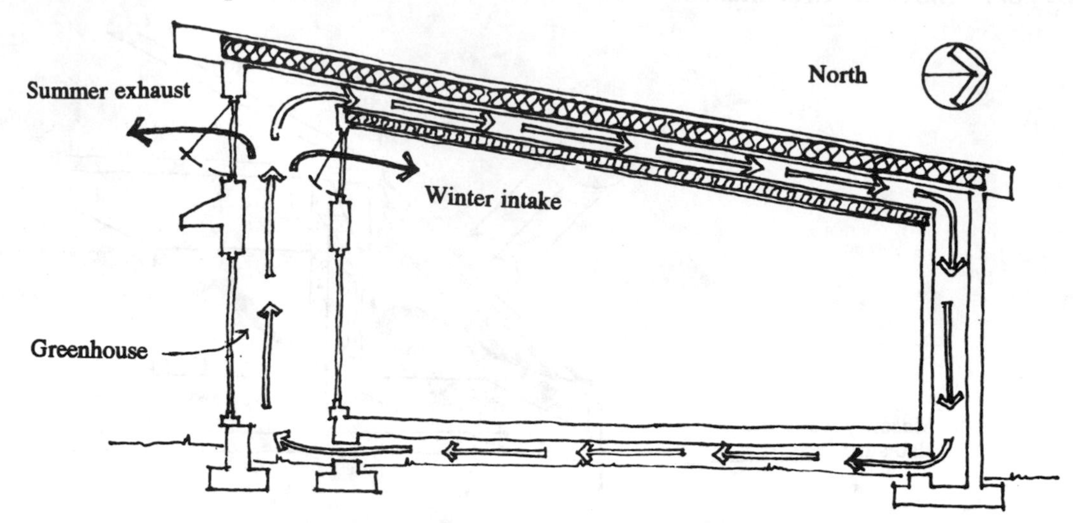

wall space, the air cools and sinks, drawing the warm air along behind it and setting up a pattern of air circulation around the living space.

At night and during cloudy days the flow reverses itself. The air in the greenhouse cools because of the loss of heat through the large glass area. This forces cool air into the crawl space on the south side, which then picks up heat from the ground that has been stored during previous sunny days. As the air warms, it flows up the north side through the attic and then back into the greenhouse.

So far, available experience indicates that the reverse air flow during night and cloudy days has not been all that strong and will probably need to be boosted by electric fans. Envelope houses do need a back-up heating system. If the house is well designed, however, this auxiliary system may be only the occasional use of a wood-burning stove.

Because the design of the house must satisfy the structural requirements of the envelope system, the architectural choices are somewhat limited.

In summary, many passive solar heating methods can and should be adopted in the construction of houses today. Simply orienting windows to the south and providing summer shade is one example. The additional cost can be small and easily made up in fuel savings. Many houses built with passive solar heating measures realize savings in heating costs that have surpassed 80%.

Active Solar Heating

An active solar heating system uses mechanical equipment such as pumps and fans to collect and distribute heat. There are two types: liquid based and air based, which designate the medium used to transfer the heat through the system.

The design of an effective active solar heating system to a particular plan is much more complicated than the application of passive solar principles. Such a design should be tackled only by an architect, engineer or another professional who has had extensive experience in this field. Active solar heating systems are not as widely used as passive systems.

Like passive systems, in those periods when the solar heat source is not available, a back-up system is needed.

Active Solar Heat for Hot Water

Independent active solar heating systems are available for your hot water, but in most cases you will need a back-up heating system. As occurs with the active solar heating for the house, the problem is one of cost, which can easily exceed $3,000 plus the back-up system. On the other hand, local conditions may permit the installation of solar hot water heating systems at a cost that can be paid back in about five years. Get some expert advice to determine whether the cost of the solar system makes this choice practical in your area.

Zoning

If your plan depends substantially upon passive and/or active solar heating, check the zoning laws for your area to make sure that further construction on adjacent property cannot block the sunshine from your lot and eliminate or reduce the capacity of your solar heating system.

Costing

The cost of installing passive solar heating measures in a house varies considerably. For example, if the only passive measure used is to face the side of the house with the most windows and doors to the south, the direct cost is about zero. If more efficient measures such as the Trombe wall or the greenhouse are adopted, the cost could run several thousand dollars. This is not an exorbitant price to pay for the one-time construction of these features, particularly if the fuel savings is high.

The expense of installing an active solar heating system can exceed $20,000 not including any back-up system.

In any case, investigate the cost tradeoffs before adopting any solar heating measure, particularly the active solar systems.

Management

The various construction items for passive heating are managed as discussed in the respective chapters. For example, changes or additions in framing are managed as indicated in Chapter 6. Because of the complexity of active solar heating systems, management during installation should be supervised by an architect or engineer experienced in these systems.

14

Heating and Cooling

This chapter provides information to assist the reader in selecting the heating and cooling system for new construction. It describes conventional central heating and cooling systems as well as auxiliary heating and cooling systems and suggests combinations of all types to meet the various requirements throughout the country.

Fuels

- *Electricity* is readily available almost anywhere. From the standpoint of the homeowner, it is clean, leaves no residue and will not contaminate the atmosphere. It requires no chimney. The cost of electricity will probably go up or down along with the cost of fuels used to generate it. One of the greatest advantages of electricity as a fuel for the future is that it can be produced from almost any other type of energy—solar, nuclear, oil, gas, coal and geothermal. For the future, relying on electricity for house energy appears to be less risky than other fuels.
- *Oil,* today, is readily available. Oil is relatively clean if the burner is regularly serviced. Storage, usually an underground tank, is needed. It does require a chimney.
- *Gas* is available in two forms, natural and liquid propane. Natural gas comes from the ground in a gaseous state, where propane gas is manufactured from crude oil. Natural gas is not available in many areas. Liquid gas is available in most areas but its cost is higher. Gas is clean and requires no

storage (except bottled gas), but it does need a chimney. It can be used to operate both heating and air-conditioning equipment.

- *Coal* was widely used for central home heating systems during the first part of the 20th century. Most localities had at that time efficient delivery systems and most homes were designed and built to provide storage space and a means of getting the coal from the truck or wagon to the storage bin. Now, most of that has disappeared. From this standpoint alone the problems associated with the use of coal to fuel central house heating systems are so great that this fuel should not be considered for this purpose. In addition, coal requires a chimney; it is dirty before and during burning; and leaves relatively large amounts of ashes which must be disposed of. Depending upon availability in your area, coal might be a good choice for fueling auxiliary or back-up heating systems such as a fireplace or a free-standing stove.
- *Wood,* like coal, is best used for back-up systems. If quantities of the hard woods such as oak, locust, birch, beech, elm and ash are available at reasonable prices, the use of wood in an energy-efficient fireplace or stove as a back-up for other systems may be a good choice. The homeowner should realize, however, that substantial labor is required to place the unburned wood in storage, to move it to the stove or fireplace as needed and then to remove the ashes after combustion.

Forced-Warm-Air Heating Systems

These systems, *with the exception of the heat pump,* are based on the provision of heat by burning fuel to create a supply of warm air that is then blown by a fan through a duct system to the various parts of the house. The air is returned to the furnace through a return duct system for reheating and redistribution.

The Heat Pump

The heat pump is one of the most innovative heating and cooling systems available today. It is also highly efficient.

The heat pump is the only heating system that does not make heat; it moves already existing heat from the outside of the house to the inside when in the heating mode and from the inside of the house to the outside when in the cooling mode (see Figure 14.1).

FIGURE 14.1 The Heat Pump

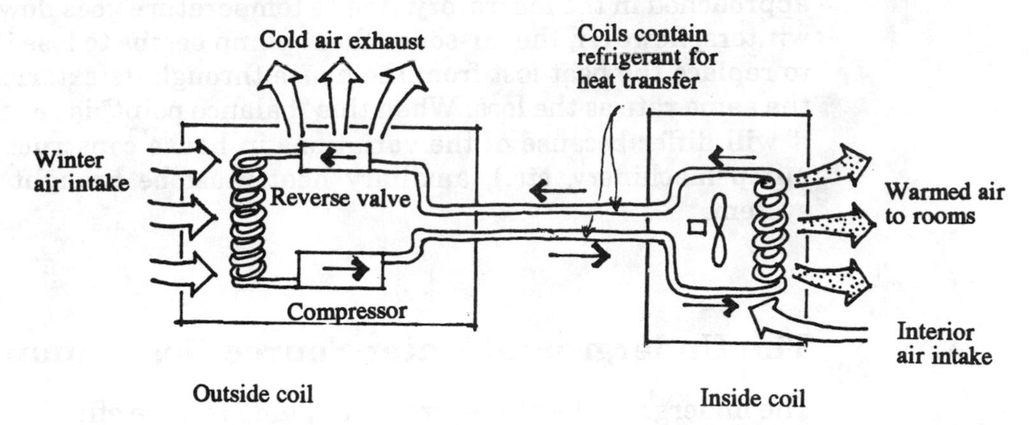

In the heating mode, liquid refrigerant moves into the outside coil, where it evaporates (expands) and absorbs heat from the outside air. The heated vapor then moves into the compressor where it is squeezed, gets hotter and is pumped into the inside coil. There the vapor condenses and gives off heat, which is picked up by the blower, added to the house air flowing through the return duct and redistributed throughout the house. For air conditioning, the cycle is reversed and the refrigerant picks up heat from the air inside the house and deposits it into the outside air.

Air-Source Heat Pump

The air-source system works in the manner described above, taking heat from the outside air in the winter and bringing it inside the house via the use of a refrigerant as a transfer medium. It can operate very effectively with outside temperatures of 30° to 25° F and above. Below that temperature its efficiency is reduced and an auxiliary heat source will be needed to maintain a comfortable level in the house. This auxiliary heat usually consists of electric resistance heat strips built into the fan coil unit. When signaled by an outdoor thermostat, these strips come on and reinforce the heat brought into the house by the refrigerant. Auxiliary heat may also consist of an oil or gas furnace installed as an integrated system with the heat pump (see page 155).

Some readers may have difficulty visualizing the extraction of heat from outside air that has reached a temperature of 25° F. All cold air contains some heat. For example, the air at 0° F has 89% as much heat as air at 100° F. Heat is completely absent from the air only at absolute zero or 460° F below zero, which can only be approached in the laboratory. As the temperature goes down in cold winters, however, the air-source heat pump begins to lose its ability to replace the heat lost from the house through its exterior skin at the same rate as the loss. When this "balance point" is reached (and it will differ because of the variations in house construction, heat pump machinery, etc.), auxiliary heat must be brought into the system.

The Underground-Water-Source Heat Pump

The underground water source heat pump is more efficient than the air-source pump. The two systems work similarly, with one very important exception. The water-source heat pump requires water such as an underground well or perhaps a lake that does not freeze during the winter. To be highly efficient, the temperature of the water source should not drop below 45° F all year round. In the United States, even in the coldest parts of the country, the temperature of underground water at depths of 40′ or more rarely drops below 45° F.

Instead of working with the outside air as a source of heat, the water-source heat pump works with the ground water through a condenser, which brings together the refrigerant in its pipe and the ground water. The net result is that the waters source pump does not lose efficiency in cold winters since it is extracting heat from 45° to 60° F water instead of 20° to 35° F cold winter air. Rarely is there a need for back-up or auxiliary heat. *Depending on the local conditions, the underground water source heat pump can do the job completely by itself.* It may be useful, however, to install auxiliary electric resistance heat strips to provide temporary heat should the water supply system break down.

In the summer, while air conditioning, the water source heat pump is putting heat from the house into water at 50° to 60° F instead of into hot outside air at 85° to 95° F. This is another bonus for the water-source heat pump, because less energy is needed to provide air conditioning.

Water-Source Requirements

The water requirements for a water-source heat pump average three gallons per minute per ton of heat pump capacity. In developing a source of water for the heat pump, care must be taken to ensure that the water discharged from the heat pump after use does not change the temperature of the water source at the point where a fresh supply is drawn to the degree that efficiency is lost. This can be prevented by establishing the discharge point of the used water at least a 100′ from the point where water is drawn into the heat pump. This solution could be very expensive, particularly if the water is drawn from a well, which could require the drilling of two lines into the well water.

An alternate and, in many cases, a less costly and more efficient system is to provide an auxiliary underground storage facility using a tank (see Figure 14.2). A new or unused concrete septic tank with the proper capacity is a good choice. Water is drawn from the well and held in the tank until drawn out by the heat pump system. After use, the water is discharged into a French well (large hole in the ground filled with rock and gravel) or a similar device that will put the water back into the ground. If you elect to go with the water-source heat pump, consult with a local contractor to determine which water supply and discharge system is best for your location and the capacities needed.

The Ground-Coupled Heat Pump

The closed loop ground-coupled heat pump is a version of the water-source heat pump. It can be installed almost anywhere. It has operated successfully in Canada, Germany, France, Italy, Denmark and Sweden. The major difference between this system and the water-source heat pump is that in the ground-coupled system the water circulates through plastic pipe underground, absorbing heat from the ground in the winter and giving off heat to the ground

FIGURE 14.2 A Water Source

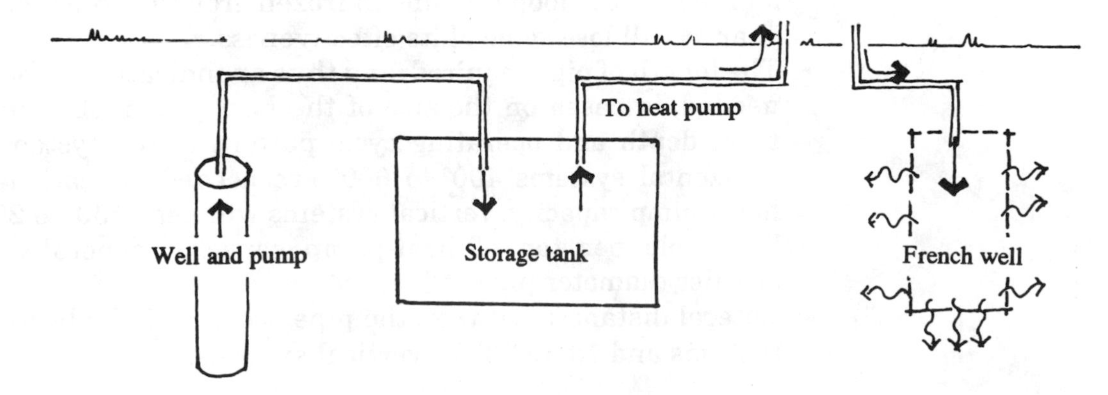

when operating in the summer as an air conditioner. Rather than continually pumping water from a well through the heat pump and then discharging it after use, the ground coupled heat system uses the same water over and over again and thus negates the problem of the large amount of water required by the water-source system and the problem of getting rid of the water after passing through the system only once.

Figure 14.3 illustrates three different forms for the ground coupling. The horizontal method places the pipe several feet underground in a horizontal plane. This is the least expensive method, because the trenching for the pipe is shallow, but it also requires the most land. If you are installing a septic system, consider laying the closed loop pipe for the heat pump in the same trenches dug for the drain field of the septic system. Moist ground gives up and extracts heat better than dry soil.

Where lots are too small for the horizontal method, there are two other choices. The deep vertical layout uses the least land of all types. Since it plunges deep, it is in contact with more stable ground temperatures and usually requires less pipe.

Where drilling is a problem (for example, when layers of rock are present) in conjunction with little land, the multiple vertical system may be the answer. With this method, several vertical pipe loops are installed at less depth than the single vertical loop.

Here are some of the details the homeowner-contractor should be aware of:

- The pipe for the closed loops should be polyethylene (PE 3408 type) or polybutylene (PB2110). In most cases, a diameter of 1½″ to 2″ is needed. Underground joints should be heat fused. This will give a life of 50 years or more to the underground closed loop.
- In the sunbelt, horizontal loops are buried from 4′ to 6′—the 6′ depth being preferred. In the north, the loop depth is 3′ to 4′ so that the sun can thaw the ground around it during the summer. If the loop remains in frozen ground throughout the year, it will lose some of its effectiveness.
- The length of pipe required, and thus an indication of the area needed, is based on the size of the heat pump, climate, soil type, depth and operating cycle pattern of the system. For horizontal systems 400′ to 600′ are needed for each ton of heat-pump capacity. Vertical systems will need 150′ to 200′ of bore hole per ton of heat-pump capacity, generally with smaller diameter pipe.
- Lateral distances between the pipe are 4′ to 5′ for horizontal systems and 15′ to 20′ for vertical systems.

FIGURE 14.3 Ground-Coupled Heat-Pump Systems

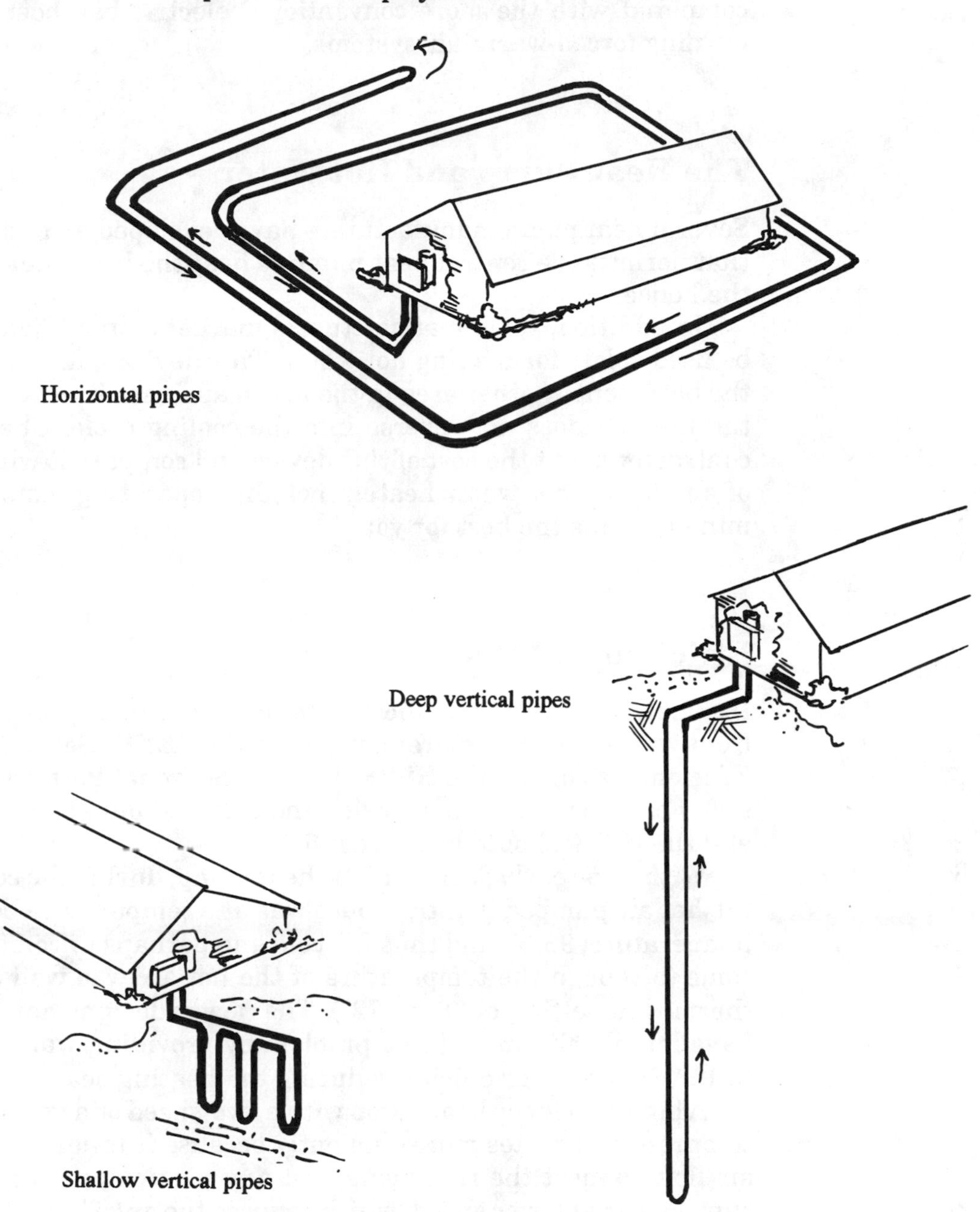

You can expect savings of up to 35% in fuel consumption cost compared with the more conventional electric baseboard and oil-burning forced-warm-air systems.

The Heat Pump and Hot Water

Several heat pump manufacturers have developed an add-on device that permits the central heat pump to heat the hot water as well as the house.

In addition, some manufacturers market a small heat pump to be used solely for making hot water. This device gathers heat from the basement or other area in the house and uses it to heat water in the tank. It does not reverse into the cooling cycle. Check with a contractor about the cost of this device and compare it with the cost of an electric hot water heater, including operating costs, to determine which is the best for you.

Efficiency of the Heat Pump

In judging the relative efficiency of different heat pumps systems from various manufacturers, compare the SEER (Seasonal Energy Efficient Ratio) and the HSPF (Heating Seasonal Performance Factor). For a high-efficiency model, the SEER should be at least 10.0 and the HSPF should be at least 6.5.

With some early designs of the heat pump, during the cold season the hot air pumped into the house was at a temperature below body temperature, 98.6°, and thus felt cold and perhaps uncomfortable to some (although the temperature of the hot air was well above the thermostat setting of 68° to 72°). The newer designs now available have largely eliminated this problem by providing warmer air (up to 15° F) at greater efficiency during the heating season.

A high-efficiency heat pump with a two-speed or a variable-speed compressor operates more efficiently because it tailors the speed of air flow to meet the requirement at any particular time. The unit runs longer at lower speed, which reduces the on/off cycling, greatly increases comfort by providing even, continuous heat in the winter and improves the reduction of unwanted humidity in the summer.

Other Forced-Warm-Air Heating Systems

Other options in your choice for forced-warm-air heating systems are the oil- or gas-fired (both natural and liquid) furnace and the electric resistance heat furnace. All three are proven and effective heating systems and each can be combined with central air conditioning. Like the heat pump, they use a duct system to transport the heated or cooled air throughout the house.

Increases in the efficiency of the oil or gas furnace can be obtained by the application of the following:

- Flue dampers, particularly if the furnace is installed in a heated area of the house with an outside supply of air for combustion
- In gas-burning furnaces, pilotless ignition, which uses an electrical spark generated only at the moment the fuel begins to flow
- Gas- and oil-burning forced-warm-air furnaces are now available with an AFVE (Annual Fuel Utilization Efficiency) in excess of 90%.

Combination of Forced-Warm-Air Systems

In many situations, a combination of a heat pump with a gas or oil furnace provides a more efficient central heating system than the single gas or oil furnace by itself (see Figure 14.4). This is particularly true in the colder climates.

This improvement is the result of using the gas or oil furnace at the colder outside temperatures, where this type of heat is more efficient; the heat pump takes over at outside temperature around 28° to 32° F and higher, where it operates most efficiently. In addition, the heat pump will provide air conditioning in the summer at no additional equipment installation cost.

The initial cost of installing gas- or oil-fired systems in combination with the heat pump will be higher than either the gas or oil system with air conditioning alone. To be economically sound, the annual savings in fuel costs should provide for a payback of this additional cost in not over seven to ten years.

Since this combination system has a low operating cost, this payback requirement should be easily met in most cases, particularly if the house is properly insulated (Chapter 12) and if anti-air-infiltration measures have been applied during construction (Chapter 11).

If you have gas available, ask your HVAC contractor about a newly developed combination hot water and space heating system that is very efficient and inexpensive to install. As heat for living space is required, the hot water heater pumps the hot water through the coils of the air heater. Air is blown over the coils, extracting heat from the water, then is forced through the duct system throughout

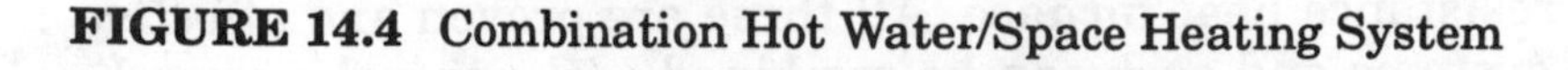

FIGURE 14.4 Combination Hot Water/Space Heating System

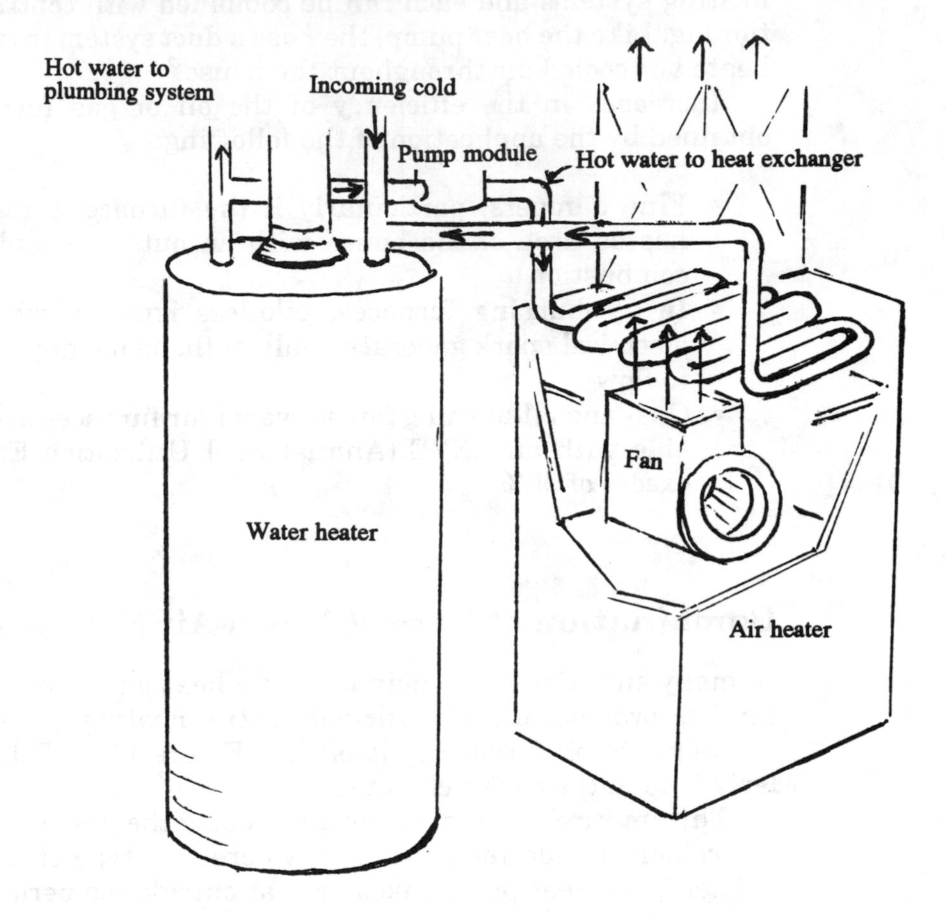

the house. After the heat is removed, the water is cycled back into the hot water heater, where it is reheated and then recycled into the air heater.

This system is easy to install, economical to operate and occupies less space than most other systems. It was originally developed for apartments, townhouses and small houses, but units are now available to heat much larger homes.

Similar systems have been developed for oil burners, but they are not as efficient as the gas burners.

Duct Work

Two materials are generally used in making the ducts for forced warm-air heating or independent air conditioning:

- *Fiberglass duct board* makes a quiet system. It is usually fabricated in the shop and assembled at the job site using staples and tape. This method of construction does not give it the rigidity nor ruggedness of the galvanized sheet metal. The fiberglass duct board should cost about the same as the galvanized sheet metal wrapped with insulation.
- *Galvanized sheet metal* is also made up in the sheet metal shop and then brought to the job for final installation. Sheet metal ducts are put together using more positive means of fastening.

By itself, a sheet metal duct tends to be noisy. This noise can be substantially reduced, however, by lining the inside of the duct with insulation material, which also provides thermal insulation. The most noise is eliminated by lining all of the duct system. Costs can be reduced by lining only selected sections that produce the most noise. Listed below in order of priority (most important are first) are those areas that should be lined:

1. First 6′ to 10′ of return duct from return air grill toward furnace
2. All of the return duct system
3. Supply ducts from furnace through the first "tee"
4. All corners

Duct Work Design and Installation Tips

A poorly designed and installed duct system will reduce the efficiency and comfort of your heating and cooling system. Consider the following tips:

- Where practical, run the duct system within the heated area. This will improve the insulation since the duct then uses the house envelope insulation rather than its own (usually less thick) and reduces heat losses through leaks in the system.
- The plan for the location of supply registers and return diffusers (entry points where the heated or cooled air enters and leaves the heated or cooled area) for any heating/cooling combination is a compromise.
- The best location of the supply registers and return diffusers in a one-story house is in the floor or low in the walls to get the most out of the heating system. Unfortunately, to maximize the air-conditioning system, both the supply registers and the return diffusers should be placed in the ceiling or high on the wall. The solution requires a compromise. Locate the supply registers in the floor along the exterior walls and in

front of windows and doors. Mount the return diffusers in the ceiling or high up on the wall in the interior of the house. With this system, the air enters the room low on the outside wall and is drawn by the duct to leave the room high and on the inside wall (or an interior hallway). Thus, the entire room is cooled or heated.

- For a two-story house, install the supply registers and return diffusers on the second floor to maximize air conditioning, i.e., both installed high on the wall or in the ceiling. For the first floor, maximize the installation for heating with the supply registers and return diffusers on the floor or low in the wall.
- Floor-mounted supply registers are more efficient than the baseboard type. It is a good idea, however, to use the baseboard register in rooms such as the bath, kitchen and laundry to prevent wax, water and other material from getting into the duct system.
- After the completion of the rough-in, require the heating contractor to install temporary covers over the outlets, both supply and return, to prevent debris, dust and other material from getting into the duct system during the remainder of the construction period.
- When installing registers in front of windows, if heavy drapes are planned for the window, mount the register far enough away from the wall so that the closed drapes will not interfere with the flow of the air supply into the room.
- When mounted outside the heated space of the house, such as in the unheated attic or crawl space, ducts should be insulated with at least 2″ of duct insulation.

Multiple Forced-Warm-Air Heating Systems

If your house is very large, you should consider a multiple system, that is, two completely independent systems with separate duct work, separate heating elements and separate cooling elements. Each of the parts of a multiple system would have less capacity than an overall single system.

Multiple systems are particularly effective in large two- or three-story houses or long ranchers. With these designs, it is difficult for a single system to distribute warm or cooled air efficiently. A multiple system solves most of these distribution problems and saves energy. In addition, the multiple system provides the basis for a zoned system, whereby the bedroom area, for instance, can be maintained at a lower temperature than the rest of the house. Even though the initial costs are higher, you will have a much more comfortable house and one that is more energy efficient.

Air Cleaners

Because of their duct network, forced-warm-air heating and air-conditioning systems are ideal for electrostatic air cleaners, which can be installed as an integral part of the HVAC system. They not only clean the air but also tend to make it smell fresher, an important consideration now that the sealing of houses reduces fresh air movement through holes in the outer shell.

Humidifiers

If your house has been constructed with few or no anti-air-infiltration measures, the air inside may be dry because of the loss of the moisture through the gaps in the exterior walls, floors and ceilings. If so, higher temperatures will be needed to gain the same comfort possible with less dry air. Should this be the case, consider the use of humidifiers.

A humidifier can be installed directly into the duct system and hooked up to a water supply so that it will continually and automatically put the needed moisture into the air.

For houses heated with systems other than forced warm air, small portable humidifiers can be placed in appropriate areas throughout the house to accomplish the same thing.

Ventilation

With the advent of efficient methods of sealing a house (Chapter 11) it is likely that the inside air may have too much moisture. Under these circumstances, rather than the movement of warm water vapor through the exterior walls, floors and ceilings, the moisture generated by the normal living activities (showering, clothes washing, cooking, etc.) will remain in the house. This lack of fresh air may make living uncomfortable. In addition, the high moisture content of the inside air may cause condensation on the inside of even dual-glazed windows and doors.

The Air-to-Air Heat Exchanger

The heat exchanger, Figure 14.5, uses electric-powered blowers to bring outside fresh, dry air into the house while moving stale, moist air out and exchanging the heat from the stale air to the fresh air in the process. Heat exchangers are about 75% to 85% efficient.

FIGURE 14.5 Schematic of the Air-to-Air Heat Exchanger

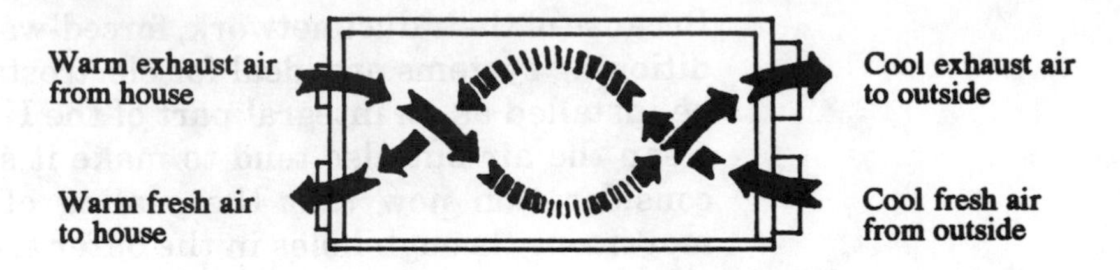

The heat exchanger should be installed in the basement or in the mechanical room along with the furnace and hot water heater. It should not be installed in an unheated crawl space, because the moisture from the stale air will freeze and the exchanger will not operate properly.

Figure 14.6 illustrates the installation of a heat exchanger system in a house that does not have a forced-air ducted heating system. Note that to operate efficiently, the exchanger must have its own duct system to extract the stale air (black arrows) from the rooms that usually generate the moisture (baths, laundry and kitchen) and to bring fresh air into the other parts of the house.

In houses with central forced-warm-air heating systems, the fresh air supplied by the exchanger should be ducted to a room where

FIGURE 14.6 Fully Ducted Air-Exchanger System

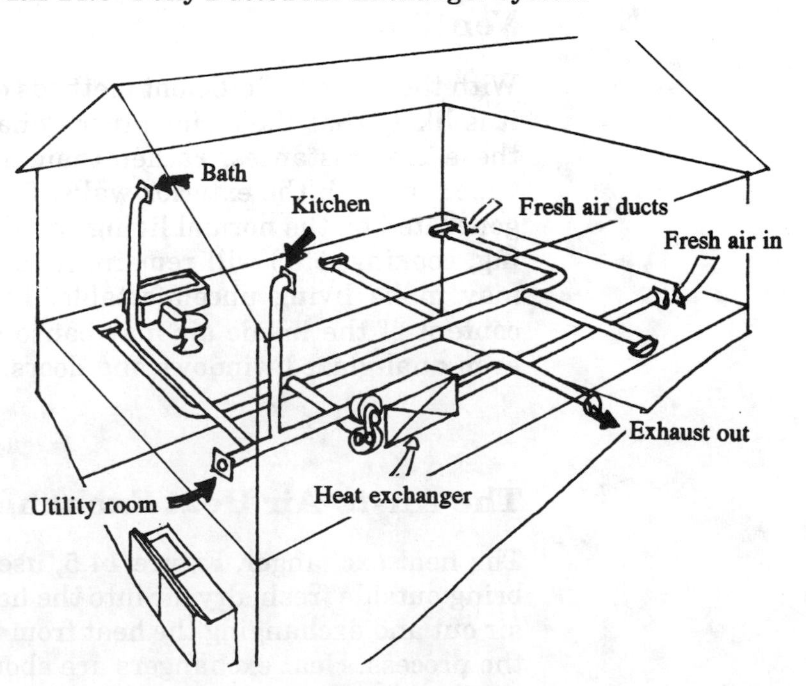

it can mix with air in the room and then be picked up by the *return* air ducts of the heating system. *There should be no positive connection between the ducts of the heat exchanger and the ducts of the central heating system.*

A heat exchanger can be installed in a heated crawl space, usually by suspending it from the floor joists. Use a flexible type of suspension, which minimizes the transfer of the operating vibrations of the exchanger to the floor system.

The rated hourly cubic capacity of the heat exchanger should be large enough to change the air inside the house once every two hours.

In very cold climates, water and ice may build up in the discharge duct of the exchanger. To monitor this problem, locate the discharge duct where it can be easily inspected and drained.

Other Heating Systems

Hot-Water Heating Systems

Although they are largely being replaced by the heating systems discussed previously, you may want to consider installing an oil- or gas-fired hot-water heating system. In this design, water is heated in a boiler by the gas or oil burner and pumped all over the house through a pipe system equipped with radiators that distribute the heat into the room.

Figure 14.7 illustrates the two-pipe system (preferred over the single-pipe system) where the supply of hot water and return of cooler water travel in separate pipes.

Note that two types of radiators are shown. One is the familiar under-the-window style modernized with an attractive grilled cover, and the other is the baseboard radiator. The disadvantages of the former are that it juts into the room and may interfere with placement of furniture and use of the drapes. If it is recessed into the wall, the remaining space available for the insulation of the exterior wall is inadequate. The baseboard system gives more outside wall coverage and thus more efficiency. For these reasons it is the preferred system.

The forced-hot-water system is a good one operating with little temperature difference between the floor and ceiling. Be sure that your system is a closed loop type with an expansion tank partly filled with air that allows the pressure to build up and raise the boiling point of the water. This permits operation of the system at higher temperatures without generating steam. Also, smaller radiators can be used.

FIGURE 14.7 Forced-Hot-Water Heating System

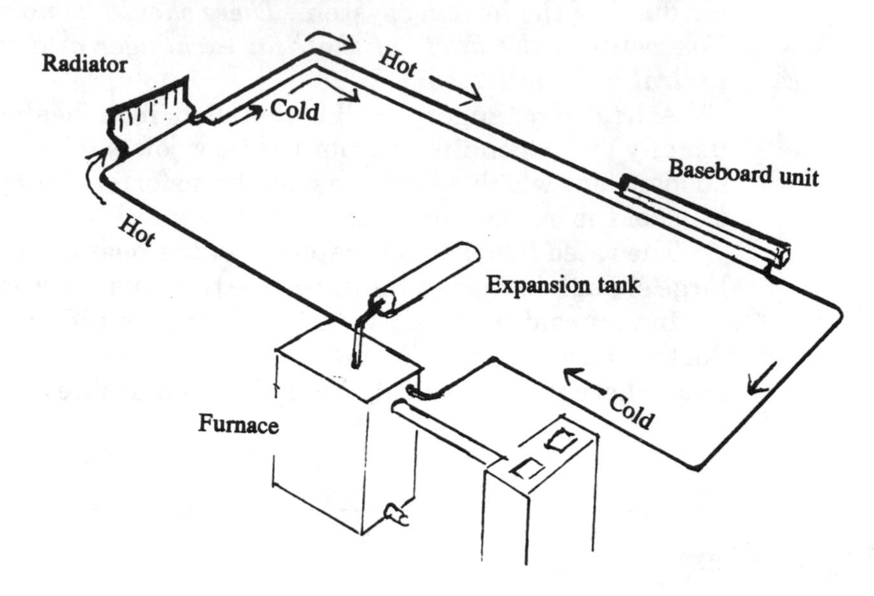

Radiant Panel Heating

Radiant heating can be attained by installing panels of hot water coils or electric resistance wiring in the ceiling, floors or walls. Design of these systems can be very tricky and should be undertaken only by one experienced in this field.

These systems depend primarily on radiation for the passage of heat. Given that radiant heat striking human skin reduces body heat loss and increases comfort, the occupant will be more comfortable at lower temperatures than with other systems.

Electric Baseboard Heat

Electric baseboard heat is one of the simplest and least expensive heating systems to install. It is unobtrusive, noiseless and, depending upon the circumstances of its use, can be very efficient. Each room has its own control permitting variations in the amount of heat provided. If these controls are properly used and if the house has been well insulated and properly sealed against air passage, the cost of operation can be low. Maintenance costs are the lowest of any heating system.

Fuel Savings with Setback Thermostats

Substantial savings in fuel can be realized by the use of a thermostat that automatically lowers the heat (setback) when the occupants of the house are away or asleep. For example, a working father and mother with two school children could have the thermostat automatically turn up to about 68° F a half hour or so before rising, then turn down to 63° F a half hour after the last person has left the house, then back up to 68 just before the first child returns from school and then once more back to down to 63 at the preset time for the night.

Figure 14.8 shows the savings from heating setbacks (thermostat setting moved down 5° or 10°) from 68° F and cooling setbacks (thermostat setting moved up by 5° or 10°) from 75° F.

Savings can also be attained when air conditioning is in operation by raising the temperature on the thermostat for those hours when the house is not occupied.

Special thermostats are available for air-source heat pumps with electric resistance back-up heat. They minimize use of the electric back-up and produce substantial energy savings.

Mechanical Air Conditioning

All of the forced-warm-air heating systems are readily adaptable at modest additional cost to include mechanical air conditioning. The

FIGURE 14.8 Fuel Saved by Thermostat* Setbacks

City	*Heating*			*Cooling*	
	A	*B*	*C*	*D*	*E*
Atlanta	15%	22%	38%	10%	16%
Dallas	15%	24%	39%	9%	16%
Houston	18%	25%	39%	9%	13%
Jacksonville	24%	30%	40%	9%	21%
Los Angeles	25%	30%	38%	11%	16%
Louisville	11%	16%	30%	10%	17%
Minneapolis	6%	10%	20%	11%	19%
Phoenix	19%	27%	38%	8%	21%

A—5° single setback (8 hours/day)
B—10° single setback (8 hours/day)
C—10° dual setback (8 hours twice/day)
D—5° single setback (8 hours/day)
E—10° single setback (8 hours/day)

*Figures are based on a Carrier "Pro-Stat" thermostat.

heat pump automatically includes it because the same machinery is used for both operations. The other forced-warm-air systems are designed to accept machinery for air conditioning using the same duct system for cool-air distribution (see remarks pages 157-58 that previously discussed the compromise necessary in designing a duct system for both heating and air conditioning).

If your primary heating system is not one requiring a duct system and you want central air conditioning, it will have to be a separate system.

Do not be excessive in oversizing the system in the hope of getting quick cooling. If oversizing is much greater than 15%, there will be a fast cooldown of air but without sufficient removal of the moisture. The result will be cold, clammy, very uncomfortable air.

The suggestions on good duct work previously discussed apply to the separate air-conditioning system as well.

If possible, locate the outdoor section of the compressor (some manufacturers make a two-section compressor) away from decks, patios, bedroom windows and dryer vents. Also avoid interior corners, which tend to accentuate the noise of the compressor. Locate the compressor in a shaded area; direct sunlight on the coils will unnecessarily increase the workload.

Auxiliary Heating and Cooling

Up to this point, the discussion of heating and cooling has been of central systems for the whole house. Many auxiliary systems available may be ideal for solving special heating and cooling problems.

Through-the-wall air conditioners, heaters and heat pumps may be the best solution to the heating and cooling of an isolated part of the house or a part that is seldom used. If the design of the house allows the mounting of this equipment through the wall, the appearance is improved and there is no loss of the use of windows. The unit is turned on only when the room is in use.

Small wall-mounted electric heaters with blowers are very effective in providing for giving quick heat to a room, such as a bath, for short time occupancy and as a supplement to the main heating system. Electric heat lamps provide similar heating.

Water evaporation coolers are very effective in hot, dry climates. They require less energy to operate than conventional air conditioners and are much less expensive. They need a more or less constant source of water.

Ceiling fans are used to provide summer comfort in areas where high humidity is not a problem. They are also useful in helping to circulate air heated by a fireplace or a stove in the winter. In houses designed with a cathedral ceiling, the ceiling fan can be used to move the warm air that may accumulate in a pocket at the apex of the ceiling.

Whole house ceiling exhaust fans are also useful in providing comfort where humidity is low. They should not be used to circulate heated air in the winter, however, since they discharge the heated air to the outside. Whole house exhaust fans are noisier than ceiling fans, the quality models of which are almost completely silent at lower speeds.

Mixtures of Heating and Cooling Systems

In many cases the best heating and cooling is provided by a mixture of more than one source. The following assumed cases illustrate this point:

Case 1: The house is designed along contemporary lines with plenty of free open space. The area has more than average sunshine throughout the year, and the summers are hot and dry. The winter has cold spells with temperatures below freezing on a few occasions. Firewood is available in ample quantities at modest cost.

For heating: Passive solar heat (see Chapter 13) maximized by efficient design. Back-up heating provided by an energy-efficient wood-burning or gas-fired fireplace. Wall-mounted electric heaters with blowers in each bathroom.

For cooling: Ceiling fans plus through-the-wall or window water-evaporation coolers.

This combination provides a very inexpensive heating and cooling system. The savings made in comparison with other, more elaborate systems, will very quickly make up for any extra cost of the construction to provide the passive solar heat. In addition to giving greater comfort during the summer, the ceiling fans will assist in the distribution of the warm air from the solar storage areas and the fireplace.

Case 2: The house design is such that distribution of heat from a fireplace would be a problem. The area has above-average sunshine, dry hot summers and mild winters. Wood is not available. Local costs of electricity are relatively low.

For heating: Passive solar heat with electric baseboard heat as back-up.

For cooling: Same as Case 1.

The electric baseboard heat is used to provide good distribution. The baseboard heat should be adjusted in each room to fit the pattern of family living. In seldom used rooms, the controls should be set at relatively low temperatures

Case 3: below-average sunshine. Wood is not readily available in quantities at reasonable cost. The area experiences very cold winters and hot muggy summers. Ground water is available.

For heating: Ground-coupled water-source heat pump with an energy-efficient gas-fired fireplace as supplement. Bathrooms equipped with supplemental wall-mounted electric heaters with blowers.

For cooling: Heat pump. Ceiling or whole house exhaust fans.

The ground-coupled water-source heat pump will provide plenty of heat by itself, but with the fireplace operating, the thermostat for the pump mounted in the hallway of the bedroom area could be kept around 65° F. In this situation, the whole house is on the cool side with the fireplace boosting warmth in high-use areas during the day. Bathrooms have electric heaters to increase the heat quickly when they are turned on. The ceiling fans will improve the distribution of the heat from the fireplace.

When some cooling is desired, the ceiling or exhaust fans used alone will improve comfort before the real heat of the summer sets in. At that time, the heat pump can be turned on.

Case 4: The area is characterized by below-average sunshine, very cold winters, no ground water, ground not suitable for ground coupled heat pump, hot summers and wood not readily available.

For heating: Air-source heat pump integrated with an oil- or gas-burning furnace forced-warm-air system. Electric wall-mounted heaters with blowers in each bathroom.

For cooling: Heat pump with ceiling or exhaust fans.

The integrated system permits both the heat pump and the oil- or gas-fired furnace to operate at their most efficient temperatures.

Case 5: The area is characterized by average sunshine, underground water source available, scarce and expensive wood, cold winters and hot muggy summers.

For heating: Passive solar heat with water source heat pump as back-up. Bathroom wall-mounted electric heaters with blowers.

For cooling: Heat pump and ceiling fan.

The purpose of presenting these different cases is to encourage the homeowner to think in terms of integrated systems for heating and cooling rather than only one system. With this approach, various methods can be used to provide heating and cooling in their most efficient mode.

An additional but very important factor to consider in designing the heating and cooling system is your living requirement. For instance, a family with four children living in a three-bedroom house with living, dining and family rooms will need more heating and cooling throughout the house than a retired couple without children living in the same house. In the latter case, large parts of the house will not be used habitually and can be maintained at lower temperatures than would be the case with the large family.

Costing

Mechanical heating/cooling systems Labor and material costs are provided by the bidding contractors. Again, specify the details so that you can be assured that all contractors are bidding on the same items. (See Appendix A.)

Auxiliary heating and cooling equipment Labor and other costs may be overlooked unless the requirements for this equipment are reviewed carefully. Most items are electrical, so the cost for labor and material should be provided by the electric contractor.

Management

In house construction the installation of plumbing, central heating and cooling systems and the electrical work are all accomplished in at least two phases. If your house is to be built on a concrete slab, the installation of these services may require three phases, i.e., work performed before the slab is poured (such as the installation of underslab ducts), work done after the framing has been completed and the final work after the wall finishes have been applied.

If your house is built over a crawl space the entire rough-in is done after the completion of the framing but before the wall insulation, sealing and drywall are applied.

For heating systems, the rough-in consists of the installation of the duct work (or piping for hot water systems) and the control wiring. The wiring to supply the power to the system is installed by the electrician.

To avoid possible conflicts in the use of space, you should arrange for a meeting on the site with your heating, plumbing and electrical contractors to go over the plans. In most cases, the plumber has the least flexibility since the waste pipe system must conform to certain minimum grades for proper flow. Should conflicts arise, it is far easier to reach a solution before work begins. This meeting also gives you the opportunity to ensure coordination among these contractors. For example, make sure that the electrician installs the power supply wiring for the hot water heater (installed by the plumber) and the electric power to the heating and cooling system.

If some of your duct work lies inside the heated area of the house (it's a good idea) and requires carpentry work after installation (for example, framing to provide a base for the drywall and/or plaster lathing), check with the carpentry crew so they can get the work done without having to make a special trip for only a two- or three-hour job. This can be costly since you may be billed for a full day's work unless the work has been clearly stated in the specifications.

If you are using a high efficiency (85%+) gas- or oil-burning heating system, make certain that your contractor is following the chimney/ventilating requirements specified by the manufacturer.

Check the duct work to ensure that:

- It has been properly hung from the framing. This is a matter of judgment. Be wary of long spans between hangers that permit sagging of the duct.
- If ducts are installed within the heated space, the drywall or plaster lathing has been hung with seams sealed on the top and sides of the duct chase area. The object is to eliminate cracks through which heat from the duct can flow and be lost into unheated areas.
- The ducts have been completely hooked up. Surprisingly, contractors do on occasion overlook the complete hookup of all sections of the duct system.
- The electrician has roughed-in to provide electric power to heating and cooling systems, including auxiliary heating and cooling such as ceiling fans. These fans can weigh over 50

pounds, so make sure that the electrical outlet can carry this weight.

- *Any of the rough-in of plumbing, heating and electrical work are completed.* The scheduling of these inspections, which are required by most local building codes, is usually done by the appropriate trade contractor. Verify completion of these inspections yourself. It can be very costly to rip out drywall and expose work for an inspection that someone forgot to schedule. And you can be assured that the drywall contractor will charge for the extra work.

15

Plumbing

The house plumbing system consists of the following parts:

- A source of water, usually a public system or a private well
- The pipeline connecting the water supply to the house
- A system of supply pipes within the house to provide hot and cold water to the various fixtures
- A waste and vent piping system to carry the waste water away from the fixtures
- A waste outlet to dispose of the waste, usually a public sewer or an on site septic system provided by the owner

Figure 15.1 illustrates, in part, a typical house plumbing system.

Supply Pipes

The homeowner has several choices, within the limits imposed by building codes, of the type of pipe to use in the water supply system:

- *Copper* makes an excellent pipe. It has a very long life, is generally not affected by corrosion and can be used to supply hot and cold water. It does have disadvantages, however. It is expensive compared with other pipe and the labor to install it is costly. Although somewhat flexible in the smaller sizes, it cannot bend to the same radius as polybutylene pipe. It can be ruptured by water freezing inside and requires special

FIGURE 15.1 Residential Plumbing System

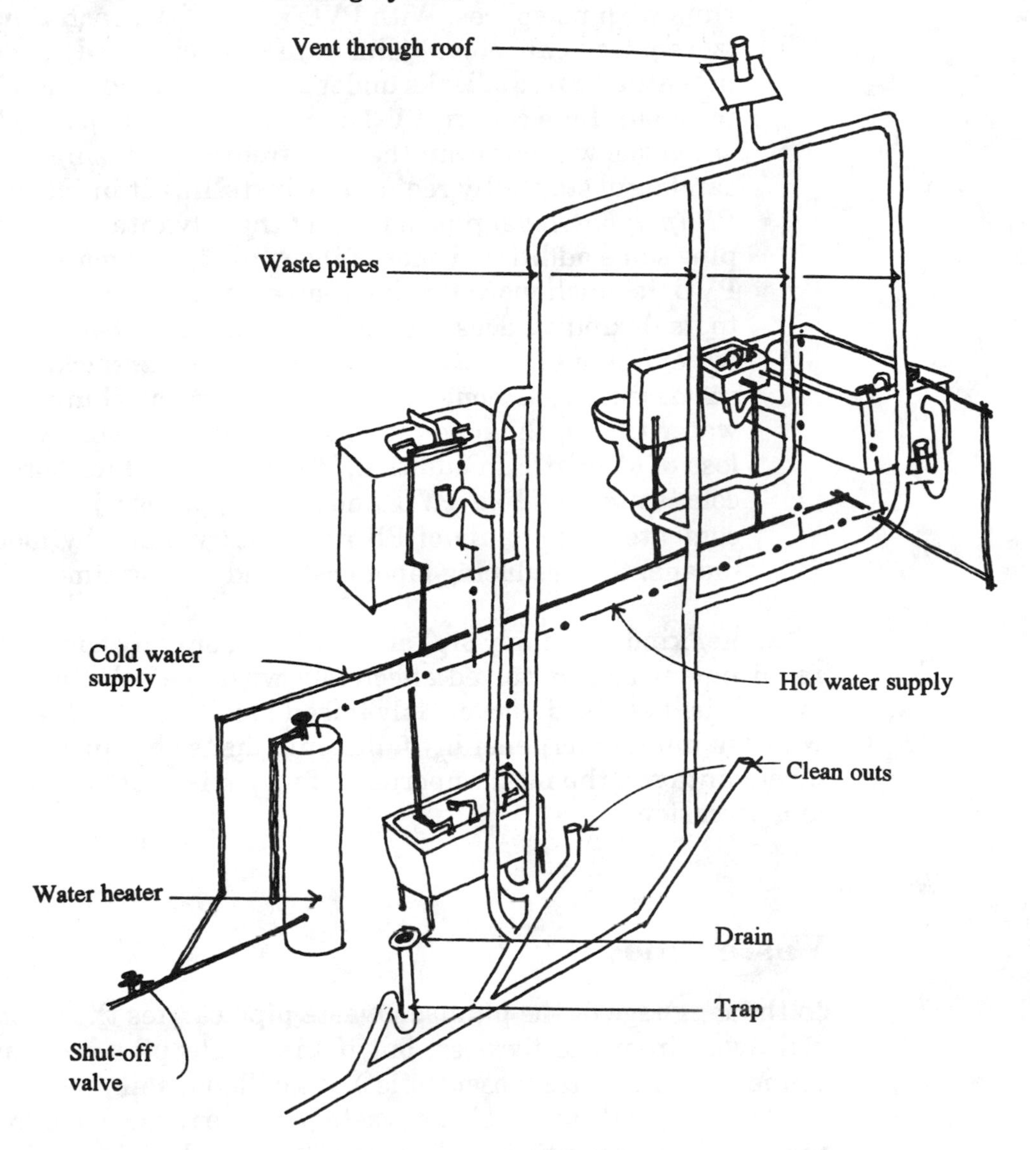

pressure chambers to eliminate water hammer (noise created by the sudden turning off of the water).

- *PVC (polyvinyl chloride) for cold water and CPVC (chlorinated polyvinyl chloride) for hot water* are two OF modern vinyl-based pipes. Lower material and labor costs make these types of pipe cheaper to install than copper pipe. They are chemically inert and therefore unaffected by corrosive material. They are rigid and lack the bending capability of copper. This shows up as a disadvantage when it is installed under or within a concrete slab, where it is most desirable to have pipe runs with no splices. With PVC and CPVC, each time a bend is required, an elbow joint must be installed, providing a potential source of leaks under the slab, which would be costly to repair. Like copper, PVC and CPVC are subject to bursting when the water within the pipe freezes. Freezing of any pipe can be substantially reduced by installing it in interior walls.

- *PB (polybutylene)* pipe has all of the advantages of the above plus some additional ones of its own. It is no more costly than PVC, is much more flexible than copper, has fewer joints due to its flexibility, does not need pressure chambers to eliminate water hammer and *it will expand and not burst when the water inside freezes,* although metal connectors may burst. When the water thaws, the pipe returns to its normal size without any loss of strength. In addition, PB can be used for both hot and cold water. Unlike PVC and CPVC, whose joints must be cemented, the joints of PB are usually made by mechanical means, thus reducing labor costs and saving time.

Other kinds of supply pipe are available, but any house plumbing requirement can be solved effectively with the selection from the materials discussed above. Galvanized steel pipe for the supply of water in homes has generally fallen into disuse because of its many disadvantages, the most important of which is short life due to rust and corrosion.

Waste Pipe

In the language of the plumber, waste pipe carries the waste material away from the fixtures, but if the waste pipe happens to be connected to a water closet (toilet), it's called a soil pipe.

The materials available for waste pipe are those listed for supply pipe, with the exception of PB, plus cast iron, which is still in general use. PB is not yet manufactured in large enough sizes for this purpose. Cast iron is less expensive than copper but more costly in both labor and material than plastic pipe.

Compared with the plastic pipe, cast iron has two advantages. It delivers the waste with much less noise; therefore, if you cannot avoid running a waste or soil pipe through a wall adjacent to the living or dining room, consider using cast iron pipe for quietness. In addition, cast iron pipe is better able to withstand the rigors of mechanical pipe-cleaning equipment.

Fixtures

Except for the fixtures, most of the plumbing system is buried inside the walls and floor of the house or in the soil. The fixtures are the items you see and use every day. Their proper selection, therefore is very important. As a general rule, try to select your fixtures from the same manufacturer. Very often a plumber can get a better price and if you want color in your fixtures, you are assured of getting a better color match.

To select your fixtures, go to one or more plumbing supply houses and take your time looking over displays. Ask questions about their features. Remember, you are going to be using these fixtures for a long time.

Tubs, Showers and Tub-Shower Combinations

Your choices for tubs are steel or cast iron base with porcelain coatings and fiberglass or other plastic. Avoid the steel tub. It chips very easily and sometimes will distort after installation and work the ceramic tile loose on a tub with tiled-shower combination.

The fiberglass tub, particularly when it's part of a one-piece tub-shower combination, has many advantages. It is easier to clean and it tends to resist mildew more than ceramic tile. Fiberglass tubs and tub-showers have drawbacks that are relatively easy to correct. They will bend under the weight of the body, and when the water is turned on and strikes the bottom of the fiberglass unit, they are noisy. Both of these faults can be largely eliminated by installing a material such as sand (usually available in ample quantities at all construction sites) on top of the plywood subfloor, then placing the tub unit on top of the sand. There should be just enough sand so that the voids in the bottom of the fiberglass unit are filled snugly.

Another possibility is a combination of a cast iron or fiberglass tub with large wall panels of patterned fiberglass or ceramic tile.

In making your choice of material, consider the ease of cleaning the fiberglass versus the greater array of colors available in the ceramic tile.

Full showers without tubs again offer a choice of fiberglass and ceramic units. Ceramic tile has a greater selection of design and an unlimited choice in size and shape. With fiberglass, you must select from the sizes made by the manufacturer. The size and shape of the ceramic-tiled shower, however, are limited only by the space available in the bathroom.

Specify that the plumbing plan includes the installation of a lead or PVC pan underneath the shower as a protection against leaks.

Two methods are used to apply ceramic tile to walls. One is the "quickset" method, where the tile is attached by a mastic to sheet rock that has been nailed to the stud walls. The other method is to place the tile in a bed of "mud" or cement that has been laid over metal lathing. The "mud" system is preferred for longer life.

Water Closets (Toilets)

Most manufacturers make water closets in the low to very high price ranges. The best buy is usually in the middle of the range. There are also available water closets that operate with less water than the standard. If water supply is a problem in your area, install this type.

Lavatories

Lavatories are wash basins. They are made of many different materials and in various sizes and shapes. The most popular are the china lavatory mounted on or dropped in a plastic-laminated vanity top, the marbleized integrated bowl and top and the Corian bowl (bowl and integrated top) made by the DuPont Company. Although the porcelain-finished cast iron lavatory is still available, it is not very popular because it requires a metal mounting ring and is very heavy and difficult to work with.

The one-piece marbleized bowl and countertop is very popular because of its appearance, ease of cleaning and low cost. The exterior coating, however, is very thin; a heavy scratch that penetrates this coating and exposes the white material beneath it is difficult to repair. On the other hand, solid plastics, such as Corian or Nevamar, are made with the color all the way through the material. If scratched, there is no color change and the scratch mark can be removed easily with steel wool or fine sandpaper. Solid plastics can be sawn and filed for an accurate fit. They are more expensive than marbleized material.

Kitchen Sinks and Laundry Tubs

Kitchen sinks are available in various sizes and are made of either stainless steel, solid plastic or porcelain-finished cast iron. The selection is one of personal preference with the option of one, two or three sections or bowls.

The more expensive line of the stainless steel sinks have more chrome alloy, which provides better looks and a reduced tendency to water spot. Otherwise, for home use there is little difference.

Porcelain sinks tend to chip when given hard usage.

Most laundry tubs are made of fiberglass. They have a deep bowl to hold more water than the average kitchen sink. If, however, the size and shape of a kitchen sink is acceptable for laundry use, it will give you an extra, more versatile sink for doing other chores. One side of this second kitchen sink can be a deep bowl and the other a shallow one.

Faucets, Shower Heads and Related Hardware

The selection of these items is a matter of personal choice, but remember that quality ones last longer.

Since it means less work for the plumber, it is to your advantage to select all of this hardware from the line of a single manufacturer. The plumbing supply house will usually help you in this regard.

Sill Cocks

A sill cock is an outside cold water tap for lawn watering and other outdoor uses. You should have at least two, and for a very large house as many as four so that your garden hose need not be excessively long and hard to handle.

In cold areas where freezing is a problem, sill cocks should be the freeze-proof variety.

Hot-Water Heaters

Hot-water heaters are available for gas, oil, electricity, heat pump and solar systems.

A standard 52-gallon water heater is normally sufficient capacity for a family of four living in a 2½ bath house. If you feel a need for additional capacity, then either go to a larger heater or change to a quick-recovery type if your fuel is electricity. The quick-recovery model has higher wattage heating elements, giving faster heating capacity.

Have the hot water heater hooked up with a timer set so that the heat comes on late at night to properly meet the demand for the following day. This arrangement will save fuel, because the heater will not automatically turn itself on every time the temperature drops a few degrees.

Hot water heaters should be located near the area of major demand to reduce the energy waste in running the water long distances. If your house has two major demand areas, install two smaller hot water heaters with one near each of these areas. The additional cost for the material will be made up by the energy saved in a few years. You will also have a hot water system that is not only more convenient to use, but also requires less water.

If you locate the hot water heater in an unheated area, add an extra coat of insulation to the sides and top (if your unit uses gas heat, do not insulate the top).

Each hot water heater should have a drain line to the outside to remove water from leaks or other mishaps.

Clothes Washers

As part of your contract, have the plumber install an in-the-wall box for the hookup faucets and drain. This box will dress up the installation and provide protection to the plaster or drywall. It is worth the additional cost.

Plumbing Plans

It is most likely that your plans contain only the location of the various fixtures. Actually, this is not a problem, since most good plumbing contractors can and will design the supply and waste pipe systems. Your plans should contain the following plumbing information:

1. Location and type of each fixture, including the manufacturer and model number for fixtures and hardware (faucet, shower head, etc.)
2. Location of water supply tap-in (If water supply is a private well, show the location of the well. If it is a public water system, show the location of the tap-in. Utility companies can supply this information.)
3. Location of the sewer-tap in (If the sewer system is to be a septic tank system on your lot, indicate the location of the tank. In most cases the local codes require that the hookup to the septic tank be made by a master plumber. If the sewer is a public system, show the location of the tap-in so that the contractor can determine the distance to lay the pipe.)

4. Your preference for pipe material

One-Supply-Pipe Plumbing System

Recently, an energy-saving plumbing system was introduced to the building trade (Figure 15.2). It offers several improvements over conventional systems.

The new system provides only one supply pipe to each fixture. The mixing of the hot and cold water is done by the valve unit located near the hot water heater. This mixing is controlled by the user selecting the appropriate button on the control panel at the fixture to get the desired temperature.

A typical lavatory would have a four button control panel with choices of *hot, warm, cold* and *off*. A tub/shower would have a panel with *warm, warm 1, warm 2* and *off* with *warm, warm 1* and *warm 2* being set at whatever temperatures the user desires. A kitchen would normally have an eight button panel: *hot, warm, cold, drink, disposal, high flow, low flow* and *off*. *Drink* provides water that has been passed through a water softener and *disposal* turns on the electricity and water supply for the disposal. *High flow* and *low flow* provide variations of flow needed for kitchen chores such as washing potatoes.

FIGURE 15.2 One-Supply-Pipe Plumbing System

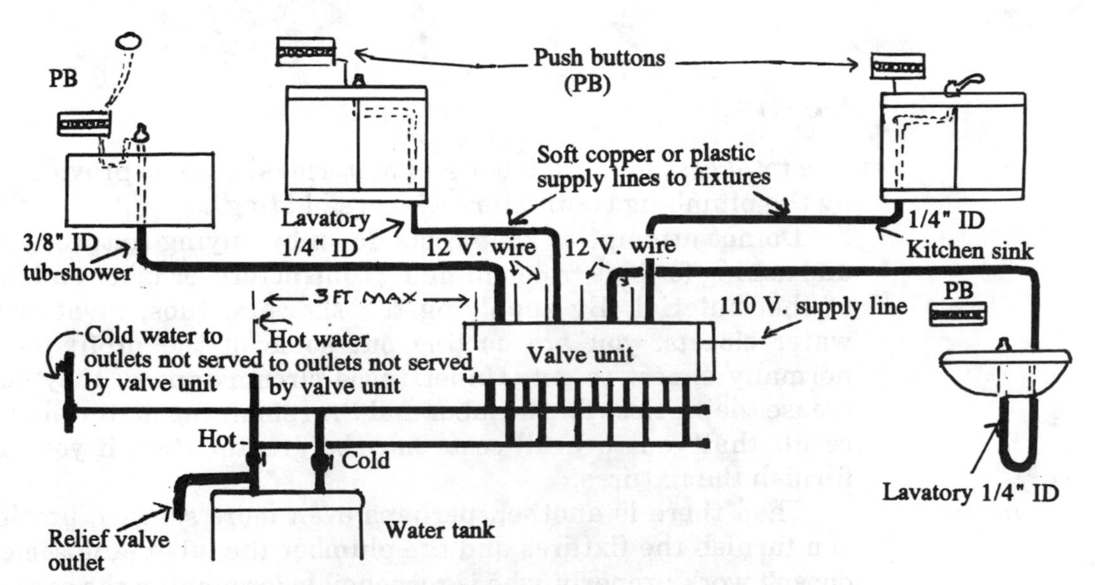

Closet bowls (toilets) and other users of only cold water are connected directly to the supply of cold water, bypassing both the water heater and the valve unit.

The supply pipes from the valve unit to each fixture are smaller than those with the conventional plumbing system—¼″ to ⅜″ rather than the ½″ to ¾″.

The advantages of the one-pipe system are:

1. It provides a substantial savings in both energy and water to operate the system. When warm or hot water is turned off by the user, the heat in the water remaining in the pipe is soon lost. When the hot or warm water is turned on again, the user typically lets it run until freshly heated water arrives at the fixture. In larger pipes, much greater quantities of water and greater amounts of energy are wasted. For example, a 50′ run of ¾″ pipe from water heater to faucet with regular plumbing holds about a gallon plus a quart, but the same run with ¼″ pipe in the one pipe system holds only a quart. Thus the conventional system will waste about five times as much water and the energy to heat it.
2. The preset temperatures are a much simpler way to mix the hot and cold water than at the faucet.
3. The preset flow will save water and reduce the amount of splashing at the fixture.
4. When a faucet is turned off, there is no pressure in the lines from the valve unit to the faucet. This eliminates the loss of water and the energy to heat it by dripping faucets. The danger of leaks in out-of-the-way areas, such as within the walls, is substantially reduced.

Costing

The total cost for both labor and material should be provided to you by the plumbing contractors who are bidding.

Do not attempt to keep costs down by buying fixtures yourself and asking for bids from plumbing contractors for labor and the rest of the material. By supplying the showers, tubs, lavatories and water closets, you are cutting out some of the profit plumbers normally expect to get. Under these circumstances, they may increase their prices for the labor and the remaining material with the result that your overall costs may be greater than if you did not furnish the fixtures.

Then there is another, perhaps even more serious, problem. If you furnish the fixtures and the plumber the labor, and something doesn't work properly, who is responsible for making the correction?

Is it the fault of the plumber's labor or did you furnish defective material? If the material is defective and has to be replaced, you have the additional problem of making the exchange and paying the plumber for removing the defective material and installing the replacement at an extra cost to the contract. On the other hand, if your contract calls for the plumber to furnish all material and labor, should anything go wrong with the system, the plumber is obliged to make the necessary corrections regardless of the cause at no change in the contract price.

In your bidding process, you may have to divide the plumbing task into three parts. First is the septic system, which should be installed by a contractor who specializes in this work. Next is the plumbing system in the house, which includes hookup to the septic system and the water supply pipe 5′ out from the foundation. The third part would be the laying of the supply pipe and the final hookup to the main. The reason for going to a specialist for the third part is that hooking up the water supply to the water main requires special equipment since the main is under pressure. Some general plumbing contractors do not have this equipment and prefer not to do this work. If you can get one contractor to do the entire job, however, it will make your management task easier.

The design and specifications for a septic system are usually provided by the local public health department. This information should be the basis for the septic system contractor's bid.

Management

- Coordinate the work of the plumbing, electrical and heating contractors as previously discussed and keep them informed when you would like to have them on the job.
- Any site work that includes excavations, such as the ditch for the water supply and the septic system, should be scheduled so that there remains at least six weeks (to allow the excavations to settle) before the final grade is done. Failure to take this precaution may result in settling of the soil after the final grading, requiring a redo of the job.
- If your lot is restrictive to the extent that machinery for the septic system cannot be brought to the septic area after the foundation has been laid, install the septic system right after the lot has been cleared and the driveway base established.
- To avoid marring the appearance of the front of the house, require that all waste and soil pipe vents exit through the roof at the rear of the house.
- To protect them from freezing, run water supply pipes through interior walls if possible. If this cannot be done, make sure

that the pipe has been run at the interior edge of the stud to allow as much room as possible for the insulation. This insulation must be installed between the pipe and the exterior wall.

- Most building codes require that after rough-in, the supply pipe system must be tested under pressure for leaks, either at the pressure of the local water system or as high as 100 pounds per square inch. The waste pipe system, which is not under pressure, is usually tested for leakage by applying a 10′ *head,* the hydraulic term for a vertical column of water. For example, a 10′ head would apply pressure to the waste pipe equal to the weight of a column of water 10′ high. If the local codes do not require this or similar testing, include it as part of your plumbing contract to be sure it is done.
- During and after the installation of the rough-in of the supply and waste pipe systems, inspect the framing and look for places where the plumber has made major cuts into the structural framing. Require reinforcement as indicated in Figure 15.3.
- *The Building Officials Code Administration (BOCA) National Building Code specifies that notches or cuts into beams, joists, rafters or studs shall not exceed one-sixth the depth of the members and shall not be located in the middle one-third of the span.*
- Your best indication of quality workmanship in plumbing is neatness. Vertical pipes should be vertical, and horizontal supply pipes should be horizontal. Horizontal waste pipe, on the other hand, must have a slope of about ¼″ for each

FIGURE 15.3 Reinforcing Studs and Joists

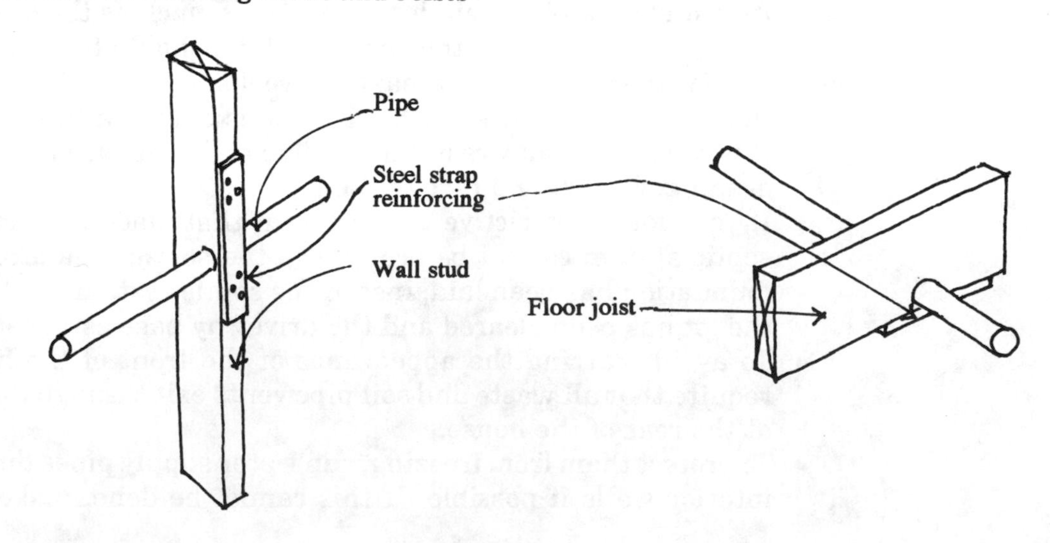

horizontal running foot. If your plumber is using PB pipe, however, this material should run directly from one point to the next, not necessarily vertical or horizontal. Joints should be clean and neat.

- If you are not using PB pipe, require shock arrester pressure chambers at all quick cut-off fixtures (faucets).
- Before scheduling the plumber for the final inspection, ensure that the following tasks have been accomplished:

  1. Vinyl, ceramic tile and wood flooring have been installed in the rooms that will receive plumbing fixtures. You will have a much neater looking floor if these types of flooring are put down first. If carpet is the selected floor covering, the underlayment should be laid but the carpet itself should be installed after the fixtures have been set. The wood floor should not be finished.
  2. All kitchen cabinets and vanities and their tops that are to receive plumbing fixtures are installed with the cutouts of the proper size already made.
  3. Appliances such as clothes washers or refrigerators with ice-making hookup that the plumber is not to furnish but is to be hooked up are on hand.
  4. All ceramic tile for showers and tub-shower combinations has been installed.

16

Electrical

The Basics of Electrical Service

To understand the discussion in this chapter and to communicate with electrical contractors, you should be familiar with a few technical terms and their use.

An *ampere* is a measure of the quantity of electricity flowing through a circuit (a wire). A *volt* is a measure of the force pushing the ampere through the circuit. A *watt* is a measure of the power in a circuit and is a result of multiplying the amperes by the voltage (for resistance-loaded equipment such as toasters, light bulbs, etc.). A circuit designed for 20 amperes working with a voltage of 110 (normal house voltage) is capable of handling 110 × 20 or 2,200 watts (resistance load). The call for power in this circuit cannot exceed this wattage or the circuit breaker protecting it will open and kill the circuit.

The number of amperes that a circuit can normally carry depends upon the size of the wire cable. The size of cable used in house construction is expressed in gauge numbers. The smaller the gauge number, the larger the wire in the cable and thus the greater the capacity of the cable to provide wattage. The actual design of each circuit should be done by a licensed master electrician.

Electrical Capacity of the House

The electrical capacity of the house is based on the input (power source), the size of the service panel and the size, number and layout of the circuits within the house.

Service Panel

The size of service panel selected should allow for future circuit and load growth. The capacity of the panel is expressed in amperes. For an average house, this rating should be at least 150 amperes; 200-ampere service is a better choice since it should be sufficient to provide extra circuitry for the future at less cost than it would be to add auxiliary panels at the time of need later. Get some help on this selection from your contractor.

Choose the circuit breaker type of panel rather than the one using the old fashioned fuse. It is much less trouble and can be reset quickly by merely throwing a lever. It will "break" again, however, if the fault in the circuit has not been corrected.

House Circuits

The time to install circuits is, of course, during construction when the structure of the house is exposed and labor costs are low. Do not underbuild in this area. The small savings could very easily be lost by having to add new circuits soon after the house is built.

Representative House Circuit Design

For a typical three-bedroom, two-bath house with a full basement, a properly designed circuit plan would include the following elements (the number represents the number of the circuit in the panel):

- General purpose circuits

1. Two bedrooms and bath (not sufficient for electric wall heaters)
2. Master bedroom and bath (not sufficient for electric wall heaters)
3. Living room outlets and kitchen lights
4. Dining room lights and hall outlets
5. Basement lights

- 20-ampere kitchen and appliance circuits

  6. Kitchen receptacles
  7. Some kitchen and all dining room outlets

- Special circuits

  8. 20-ampere laundry appliance (irons, etc.)
  9. 15-ampere circuit for fuel-fired furnace
  10. and 12. 240-volt circuit for central A/C or heat pump (Note that the 240-volt requirements take up two circuits in the service panel.)
  11. 20-ampere workshop circuit

  13 and 15. 120–240-volt range circuit
  14 and 16. 240-volt hot water heater circuit
  17 and 19. 240-volt circuit for clothes dryer
  21. 20-ampere circuit for dishwasher
  22. 20-ampere circuit for garbage disposer
  18, 20, 23 and 24. Spares

This listing of circuits is not an attempt to design the electrical circuitry for your house. Rather, it is an attempt to show you the magnitude of circuits for a typical house so you are able to judge whether or not your electrical circuitry plan is in the ballpark.

The Electrical Plan

The blueprints for your house may not include a complete electrical plan, or the plan may not suit your needs. As in plumbing, however, this is not a serious problem. You should review the plan and be sure that it includes the following:

1. Location, identification and power requirements for each major appliance such as the dishwasher, clothes dryer, clothes washer, stove and so forth. Select these appliances by model number and manufacturer from the appliance supplier who can provide the necessary electrical data.
2. Location on the blueprint of all wall receptacles, wall switches and overhead lighting that you want. (If your blueprints already include a more or less complete electrical plan, make the changes to satisfy your preference.)
3. The size of the service panel. (You will probably have to get some professional help from an electrician for this.)
4. Manufacturer and model number of all electrical fixtures that you expect the electrician to supply as part of the bid.

arrangement provides a quick source of heat for rooms that are seldom used throughout the day and permits the overall temperature in the bedroom part of the house to be kept low with a resultant savings in fuel.

- Specify 20-ampere circuits as a minimum for most outlets.
- Wall switches located at the room entry should consist of the switch itself plus an outlet for plugging in vacuum cleaners, floor waxers and other house cleaning equipment. This arrangement avoids the problem of finding a low wall switch (usually behind furniture) to power electrical equipment for temporary use.
- Avoid recessed lights in ceilings with unheated space above. They cannot be properly insulated and they leak air badly. If recessed lights are an important part of your design, consider installing a dropped ceiling (one which is suspended a foot or so below the regular foot ceiling). With this arrangement, ample insulation and anti-air-infiltration sealing can be accomplished above the regular ceiling.
- Include lights in all closets controlled by a wall switch mounted on the room side. Even shallow closets with bifold doors will need lighting on occasion. Lights operated by pull chains often malfunction.
- Specify that in all exterior walls, cables installed horizontally will run along the top of the bottom plate. This will interfere the least with the effectiveness of the wall insulation. Most electric cable is sufficiently flexible so that this requirement should pose no problem.
- Include some outside electrical outlets—at least two on up to four for large ranchers. They should be equipped with ground fault circuit interrupters for additional protection from the hazard of wet ground.
- Consider some outside spotlights or floodlights to facilitate use of the yard and provide some security.
- Install a ceiling outlet in the garage for an automatic garage door opener, whether or not you have this device now. You will probably want to add it later and installing the outlet now will be less costly.

- *Don't forget the following:*

  1. Television antenna wiring
  2. Wiring for central vacuum systems
  3. Wiring for smoke and fire alarms
  4. Burglar alarm systems
  5. Intercoms
  6. Telephone wiring—Be liberal in the number of outlets

Copper or Aluminum Wiring?

Copper and aluminum are both efficient conductors of electricity. Unfortunately a few years ago aluminum wiring got a very bad reputation for causing fires and other hazards in homes. If the hardware, such as connectors, lugs and terminal blocks, is designed specifically for aluminum wiring and properly installed, there is no reason why aluminum wiring should not be used.

Although aluminum wiring is less expensive than copper, the savings is relatively small. For the greater capacity circuits requiring #8 cable or larger, aluminum is used as a common practice.

Lighting

According to the Department of Energy, 20% of all the electrical energy produced in this country is used for lighting homes and businesses. Furthermore, a reduction of 20% to 50% of this power use could be made primarily through the use of more efficient lamps (bulbs) available on the market today.

Interior House Lighting

In selecting the bulbs for lighting your house consider the following points:

- Incandescent bulbs are by far the most widely used. *They are also the most inefficient.* About 90% of the energy consumed by an incandescent bulb is dissipated in the form of heat. The efficiency of bulbs increases as the wattage increases. For example, a 100-watt incandescent bulb produces the same amount of light as two 60-watt bulbs but uses less electricity.
- Incandescent bulbs have a short life span and near the end of this span the light out put will decrease as much as 20%.
- "Long life" incandescent bulbs are the least efficient. They should be used only in areas where an incandescent bulb must be used and replacement of a burnt-out bulb is difficult.
- Most devices on the market designed to prolong the life of an incandescent bulb reduce both the light output and the efficiency of the bulb.
- A tinted bulb has a lower light output than a standard incandescent.
- Fluorescent bulbs are up to five times more efficient than incandescent bulbs.

- Fluorescent bulbs are available as long, narrow, U-shaped and circular tubes.
- Adapters are available to convert an incandescent socket to take a fluorescent bulb.
- Although the initial cost of fluorescent bulbs is higher than that for incandescent bulbs, this additional cost is made up by the greater efficiency and longer life of the fluorescent.
- Fluorescent bulbs are available in several shades of white and some will blend with incandescent bulbs.
- A 40-watt fluorescent bulb is more efficient than smaller ones.
- Straight fluorescent bulbs are more efficient than circular ones.
- Contrary to popular belief, to save electricity fluorescent bulbs should be turned off when not in use, even for only a few minutes.

Outdoor Lighting

Illumination of wall surfaces, doorways, shrubbery and walkways can be accomplished with low-wattage incandescent bulbs. If your plans call for the lighting of large yard areas, however, consider using the very efficient mercury lamp. It is available in sizes ranging from 40 to 1,000 watts, the 175-watt size being equivalent to about 300 watts of incandescent lighting. This lamp has one drawback in that it has a start-up delay of from one to seven minutes from the time it is turned on until full illumination has been reached.

In designing your outdoor lighting system ensure that it doesn't infringe upon your neighbors' privacy. Also, position exterior lights so that they do not shine in your face as you go out the door. This dangerous situation may cause anyone using the doorway to trip or fall.

Light Fixtures

Light fixtures are available in a wide variety of prices and, unless you have a well thought out plan for their installation, it is best to lean toward the inexpensive fixtures with the idea of replacing them some time in the future when you have had time to decide exactly what you want. As a suggestion, select good-quality fixtures for the entrance foyer, dining room and baths. Most other fixtures, especially those in the bedrooms, are very seldom noticed and inexpensive fixtures would probably be satisfactory. A final selection can be made more easily after the house has been furnished and lived in for a while.

Costing

- The provision of all labor and material, including electrical fixtures, is normally included in the bid of the electrician. In addition, this bid will include the labor and materials for the electrical hookup of plumbing, heating and air-conditioning units requiring electric power. If your plans call for electric-baseboard heat, the electrician will furnish the material and labor rather than the heating contractor. In any event, your specifications should be clear about the work to be bid on. Usually, the electrical contractor does not provide appliances but will hook them up.
- One easily forgotten electrical item is the kitchen exhaust fan. If it consists of a through-the-wall fan only, the electrician should furnish the material and the labor. If it is a range hood, the material is normally provided by the appliance supplier with the electrician doing installation. If duct work is necessary to the outside for the fan, most electricians will do this work, but make sure to detail it in the specifications.
- The electrical contractor furnishes all labor and materials for the temporary electrical service.
- In the preparation of their bids for the electrical work, ask each bidder to describe the number and type of circuits to be provided. You can make a rough comparison and determine which is the better bid from the standpoint of circuitry.

Management

- Arrange for the installation of temporary electric service early. The service is needed by the carpentry crew, and getting the approvals from the power company and the local authorities may take some time. Without this power in place, the framing work will be delayed.
- Have the telephone wiring installed at the same time as the electrical rough-in. Most electricians will do this work.
- Coordinate the work with the plumbing and heating contractors.
- Inspect the rough-in for the following:

    1. That junction boxes for all switches, outlets and fixtures have been installed according to plan

2. That power has been provided for all major appliances and plumbing and heating elements
3. That lateral cable runs in exterior walls are along the top of the bottom plate where practical so as not to interfere with the wall insulation
4. That the electrical inspection of the rough-in has been made before the insulation and drywall are installed
5. That ceiling outlets will hold the weight of fixtures
6. That your electrical fixtures are selected and ordered early (Some of them may have a lead time of several weeks or months. Check this with the supplier and if unusually long lead times exist, select substitute items.)
7. That the location of the junction boxes for wall switches near doors allows enough clearance to install the door trim without interfering with the switch cover plate
8. That the swing of the doors and box locations insure that the switch will not end up behind the door when open. In most codes this is not acceptable.

- Schedule the electrical contractor's final installation when the following have been completed:

    1. Interior trim has been completed and all cabinets installed.
    2. The plumber has installed all plumbing equipment that must be hooked to electrical power by the electrician.
    3. The heating contractor has installed all equipment that requires electrical power.
    4. All major appliances requiring electrical hookup are on hand and have been installed by the plumber where required.
    5. All electrical fixtures are on hand. In the event that one or two are missing with still some lead time, the final electrical inspection can be scheduled by using temporary fixtures in place of those that have not arrived. This will permit the issuance of an occupancy permit (assuming that all other work is satisfactory) and you can move into the house without waiting many days for a couple of light fixtures.

If your plans include a central vacuum system, ensure that the rough-in for this system, including the tubing, is installed by either the electrical contractor or a separate vacuum contractor before the wall is covered with the drywall or other finish.

17

Interior Wall Finishes

Conventional Plaster

There are two conventional plaster systems: two-coat and three-coat. The two-coat system can be applied over plaster board lathing or masonry, but not metal lathing. It consists of a base coat, which is doubled back for additional thickness and strength, followed by the finish coat applied after the base has thoroughly set. The finish coat can be smooth or textured. The minimum total plaster thickness should be ½″ over gypsum lath and ⅝″ over masonry.

The three-coat system can be applied over metal lathing, plaster board lathing or masonry with a minimum plaster thickness of ⅝″. It consists of a first or scratch coat followed by a brown coat and then the finish coat which is either smooth or textured.

Either of these systems will provide a fine white wall finish, the three-coat method being preferred because it produces a stronger wall. It is also the most expensive alternative. Although it is the intent of the manufacturer of the plaster materials that finished walls be decorated, neither of these two plaster finishes need be painted or papered.

Compared with other wall finishes discussed in this chapter, conventional plaster is costly in time and labor. Its application is a craft that requires a high level of skill. On a large house it can take as long as two or three days just to apply the scratch coat and allow drying time before the brown or second coat can be put on. During the entire plastering process, because of the large array of saw horses and work platforms needed, almost no other interior work can be done.

Drywall

The most widely used wall finish is drywall (also called sheetrock and gypsum board). It is also the least expensive choice.

Drywall must be painted because its finished appearance is not uniform. All joints are more or less white, whereas the rest of the area is buff, the color of the surface paper.

Three attachment devices are used in hanging drywall panels to the wall studs and the ceiling joists: adhesive, drywall screws and annular ring drywall nails. The principal problem to avoid is the popping of the nails caused by the shrinking of the studs and joists as they begin to dry out. For this reason, avoid using nails. Although the drywall screw is the most widely used method of hanging, the best system is to apply adhesive made for this purpose to the studs and the joists with the use of enough screws to hold the drywall panel in place until the adhesive bonds. Screws on the side wall are put around the edges of the drywall and will be covered with the wood baseboard, crown molding or a taped finished seam. In the ceiling, the need for additional support in the center of the drywall panel is met by installing screws temporarily down the middle of the panel at every other joist. Once the adhesive has bonded, these screws are removed and the holes filled during the finishing process.

Panel adhesive cannot be used as a method of hanging exterior walls or ceilings if the vapor barrier consists of polyethylene sheet material nailed to the studs and joists. See Chapter 13, "Insulation." The polyethylene will prevent the adhesive from bonding to the studs. An excellent alternative is to install foil-backed drywall panels with the adhesive. The foil provides a very good vapor barrier, and you save on labor costs since the labor to apply the polyethylene has been eliminated.

The taping and finishing of the joints of the drywall are accomplished in three phases. In the first phase, joint reinforcement tape is set in joint compound over each joint. A second and wider coat of joint compound is applied after the first has dried and finally a third, still wider coat is applied. This process requires at least three days. More time may be needed if the air has a high moisture content and the drying is slow.

Drywall panels are manufactured in sizes of 4′×8′, 4′×10′ and 4′×12′. Most work is done using the 4′×12′ size hung horizontally on the walls and perpendicular to the joists on the ceiling. Although the larger panels are heavier and more cumbersome, their use will eliminate joints created if the smaller sizes are used.

Except for fire-rated drywall mentioned below, the normal thickness is ½″. It offers the best compromise between sound proofing, fire resistance, rigidity of walls, ease of installation and cost.

If you intend to finish the ceiling by spraying textured material on the gypsum panels, to avoid sagging, use ⅝″ panels in the ceiling if joists are installed 24″ on centers (the usual case when prefabricated roof trusses are used in the framing). If the ceiling joists are installed 16″ on centers, the ½″ thick panel will provide adequate strength to hold the extra weight of the heavy textured finishing material.

For rooms such as baths, kitchens and laundries, a special water-resistant drywall should be used, called "green board" because of the green color of the face paper. It is finished in the same manner as standard drywall paneling, but *make sure that the end joints are sealed with water-resistant sealer,* a process often overlooked in properly installing this type of material.

Building codes usually require a ⅝″ thick fire-resistant drywall panel on walls adjacent to potential fire hazard areas such as attached garages and furnace rooms. Even if not required by the local code, its use is warranted for the safety gained at very little additional cost.

Fiberboard

An alternative to drywall is fiberboard. It is made from gypsum ore and recycled paper, mainly newsprint. Fiberboard is stronger, easier to install, more fire resistant and controls sound and moisture better than drywall. It is also heavier, harder to score (cut for breaking the board so it will fit) and may cost more than drywall.

Plaster Veneer

Plaster veneer is available in either a one-coat or a two-coat system. Both are applied over a gypsum base or plaster lathing. Joints are taped with a fiberglass mesh tape and stapled to the ends and the edges of the gypsum before the plaster is put on.

The one-coat system consists of a special finish plaster applied as a thin base coat and then immediately "doubled back" for increased strength. The total thickness of the plaster is about 1⁄16″. The finish is rough like sand and white in color.

The two-coat system consists of the application of a thin base coat 1⁄16″ to 3⁄32″ thick and the immediate addition of a finish coat for a full plaster thickness of ⅛″. The texture of the finish can be rough floated or smooth troweled. The color is white.

Both of these plaster systems provide a very hard finish, which is much less susceptible to damage than drywall paneling and conventional plaster. It also has greater strength than the drywall. They can be applied in a matter of two or three days, compared with the much longer time required to complete conventional plaster.

In comparing the cost of labor, the time to do the job and the finished product, plaster veneer offers an excellent choice for interior wall finishes. If you select this system, it is very important that the contractor has the experience and skills to do the job properly. Plaster veneer should not be attempted by amateurs.

Closet and Garage Finishes

In some areas of the country it is common practice to finish the interior of garages and closets (assuming a drywall selection) with only one coat of the joint compound instead of the usual three. This small savings is of doubtful value and the outcome is a very poor looking finish.

Paneling

Wood paneling comes in many different varieties, both real and artificial (see Figure 17.1). A line of prefinished paneling giving the effect of wallpaper is also available. Paneling has the advantage of low maintenance, but little flexibility in changing the decorating scheme of a room several years later.

Paneling should be installed over a base of finished or unfinished drywall to provide better sound proofing. If the framing of the wall consists of studs on 2′ centers or if the paneling is thin, it is even more important to apply the paneling over a drywall base to give needed strength.

Paneling requires its own matching trim material for inside corners, outside corners and crown molding at the juncture of the paneling and the ceiling. The base used at the juncture of the paneling and the floor can also be matching trim or it can be the same baseboard used throughout the house. The paneling supplier usually carries all the trim needed to match.

Paneling is best installed by using adhesives and a few matching nails to hold the panel in place while the adhesive sets. These nails can be found at the paneling supplier.

FIGURE 17.1 Wood Paneling and Trim

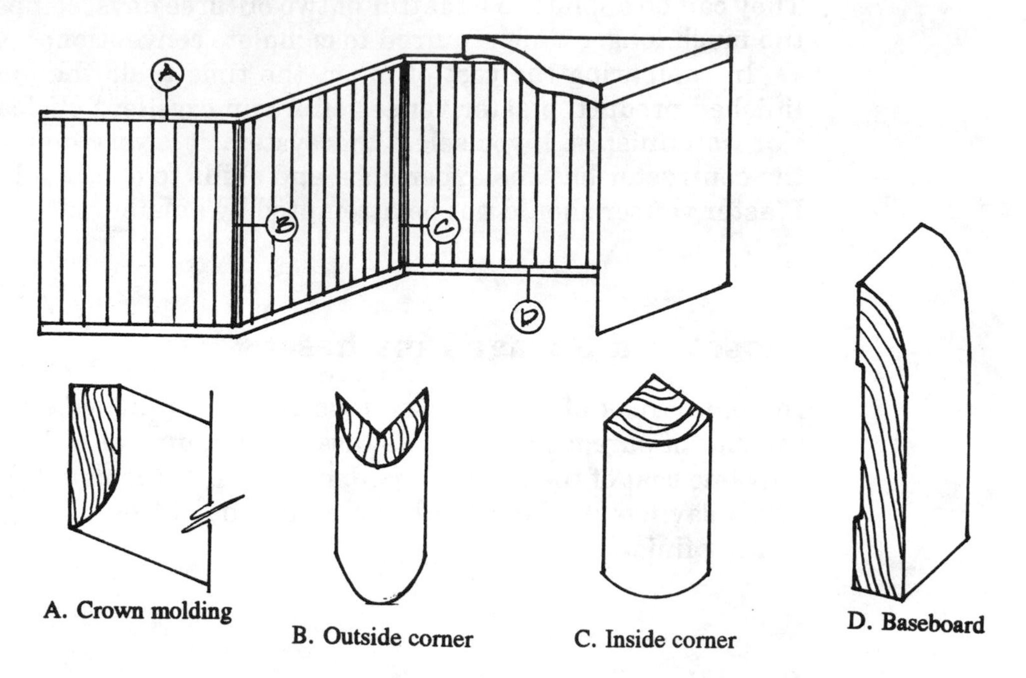

Interior Wood Trim

Wood trim gives the interior of the house a finished appearance. Its selection and installation is important since it is constantly in view and thus contributes to the quality of the house (or lack of it, if done poorly). Figure 17.2 illustrates various types of interior trim and the application of each.

- *Crown molding* is applied at the juncture of the ceiling and the walls. It is made in several sizes with the larger ones being used in larger rooms or in rooms with higher than standard ceilings (more than 8′) or both. Crown molding can also be made up by using two or more pieces of trim to give a more massive appearance.
- *Casing* is the molding applied around the doors and windows to cover the juncture of the door or window frame and wall finish. It is available in several sizes and styles; the ranch style is normally used in the contemporary houses. In addition to casing, the window trim requires a sill or stool and trim beneath the stool (an apron). Alternatively, the window can be trimmed with the picture frame technique, where the casing is also used in place of the stool and apron.

FIGURE 17.2 Typical Wood Trim

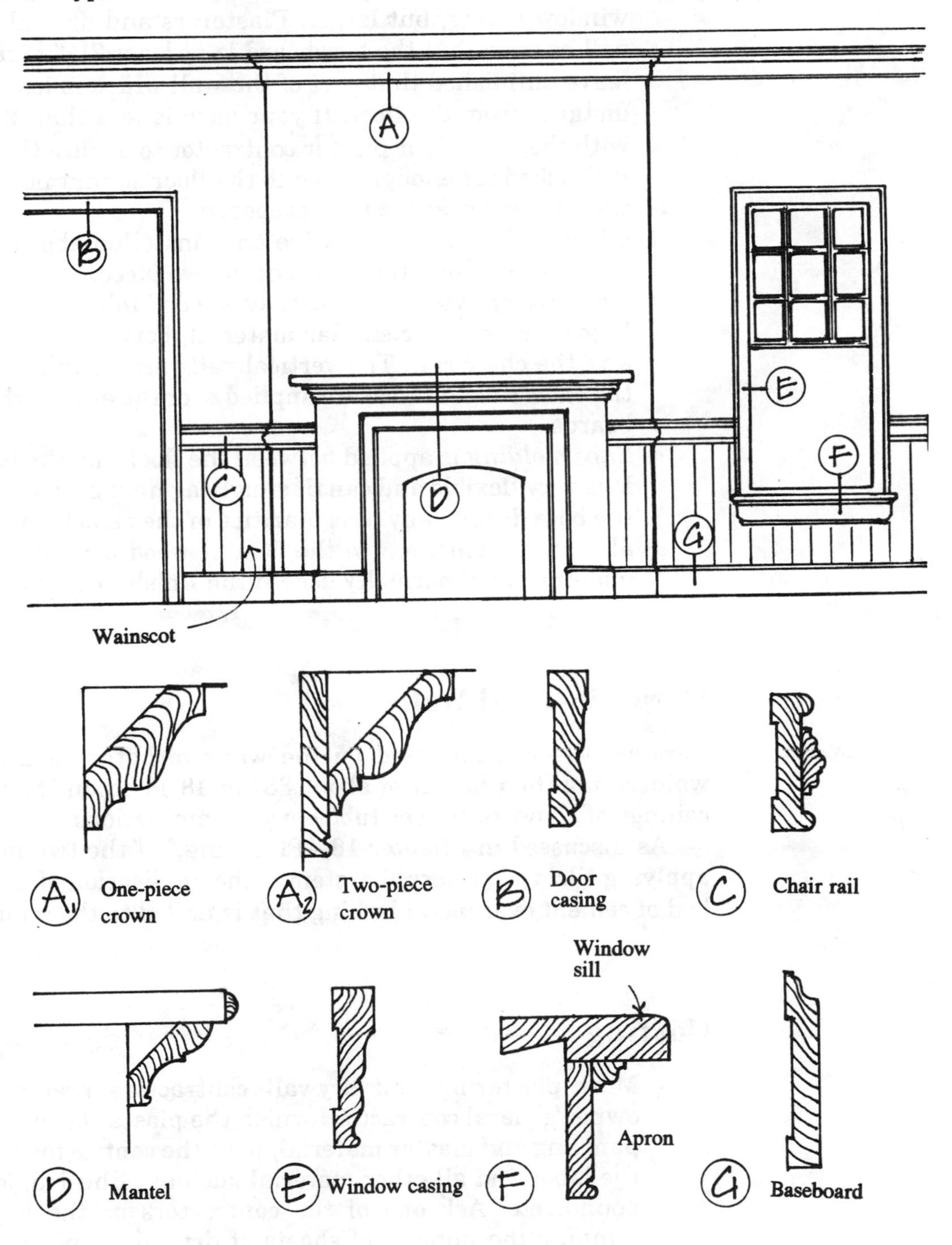

- *Base* or *baseboard* trim covers the gap between the floor and the wall finish. Its style is usually the same as the door and window casing, but larger. Plasterers and drywall finishers will assume that the baseboard is at least 2½″ high and will leave unfinished that part of the wall which is less than that distance from the floor. If your base is less than 2½″, check with the drywall or plaster contractor to ensure that the wall is finished far enough down to the floor so that no unfinished wall will show above the baseboard.
- *Chair railing* is a decorative trim installed about 32″ above the finished floor. It can be one or two piece.
- *Wainscoting* is the application of wood planks with tongue and V–groove, or other similar material, between the baseboard and the chair rail. The vertical rails are installed first, and the chair rail and base are applied over the ends of the vertical boards.
- *Shoe molding* is applied between the floor and the baseboard. It is very flexible and can fit snugly against both the floor and the base despite any irregularities of the structure, which are always present even in the best of wood construction. Shoe molding is not normally used if the finish floor is carpeting.

Ceramic-Tiled Walls

Ceramic tile is often used on the walls of baths in the form of wainscoting (to a height of about 38″ to 48″) and on the walls and ceilings of showers or over tub/shower combinations.

As discussed in Chapter 16, "Plumbing," of the two methods of applying tile, the preferred system is the application of the tile in a bed of cement over metal lathing that is nailed to the studs.

Costing

- Most plastering and drywall contractors prefer that the owner/general contractor furnish the plaster lathing, drywall paneling and plaster material, with the contractor furnishing the labor and all other material such as adhesive, joint compound, etc. Ask one of the contractors or the supplier to compute the number of sheets of drywall or plaster lathing needed.
- The cost of paneling is provided by the supplier; the trim crew does the labor. Do not forget the special trim material to match the paneling.

- The cost of the interior wood trim is computed by the supplier. The labor cost is provided by the trim carpentry crew.
- The cost of all material and labor for the ceramic tile is included in the tile contractor's bid, who must understand the method of tile application you want.

Management

- As mentioned in Chapter 6, "Framing," you may need to remove or delay the installation of a window so that the drywall or plaster lathing can be put into the house with the least additional labor for spreading it around the rooms. Ask the supplier to visit the job site and suggest the best method to use.
- Be certain that if the drywall is moved into the house before the completion of all required inspections (framing and electrical, plumbing and heating rough-in), the material does not block the view of the inspectors. Or, better yet, don't order the drywall or lathing until all inspections have been completed.
- Do not schedule any other work inside the house until the drywall or plaster work has been completed.
- Wood paneling should be installed after the plaster or drywall work has been finished. It is part of the trim carpentry job.
- If you are not sure as to exactly what features of trim work you want other than the required window and door casing and the base molding, ask the trim carpenter to show you a sample of a chair rail, crown molding or any other features you might like.
- If you have long runs of trim, for example in crown molding, that require splicing, the spliced joint should be cut on an angle and glued as well as nailed. Corner joints should be made so that the wood can expand and contract without the parting of the joint in clear view. Most trim carpenters will know how to use the coping saw to accomplish this.
- One of the first jobs the trim crew will do is to hang the interior doors. Make certain that this material (including hardware—hinges, knobs, etc.) as well as the rest of the trim material is on hand at the proper time so that once the trim carpentry begins, the crew can work straight through with no stops due to your failure to order the material on time.
- Your doors can be prehung or not. Since prehung doors require less labor by the trim carpenters, make sure that this point is listed in your bidding information.

- The trim job includes the installation of towel racks, medicine chests, paper holders and other similar items. Be sure that they are on hand.
- Check your plans carefully to be certain that the doors of the medicine cabinets swing properly, that is, left hand hinge or right hand depending on placement of the cabinet.

18

Cabinets

Manufacturing Methods

Most cabinets are manufactured in one of the following three ways:

- *Custom* made in a local cabinet shop according to your exact plans. This method offers an almost unlimited selection of material, design and finish. Custom cabinets are also the most expensive type. If your needs are unusual, this may be your best choice. If possible, get bids from several cabinetmakers before making your final selection.
- *Factory* made to the specifications of the lines the factory carries. Your choices of design, material and finish are limited by these standard factory lines. Cabinets are made up in advance and stocked by the factory or by outlets for the factory. Costs range from very expensive to least expensive of any line depending on the individual manufacturer.
- *Custom factory* cabinets made at a centralized plant. Your cabinets are not built until your order has been received by the plant. You can expect a much greater choice in design, material and finish than most factory-made cabinets. Prices are usually less than custom-made cabinets and more than the medium and lower lines of the factory-made cabinet.

Source of Cabinets

A cabinet store usually offers the greatest selection of cabinets. This supplier usually has contacts with several factory lines, including custom factory, as well as with one or more custom-cabinet shops. In addition to offering a full range of cabinet choices, this supplier is usually able to provide good advice on design of your kitchen.

The custom-cabinet shop can supply cabinets made in its shop built to your specifications. Some of these shops also carry sidelines of factory cabinets.

The building supply store usually sells factory-made cabinets and you provide the installation. Most carpenters who are qualified for interior trim work are also able to install cabinets and counter tops. If you buy cabinets from a supplier who does not install them, be certain that your trim carpenter has the experience to do so and that the cost of this labor is included in the bid.

Types of Cabinets

Unfinished wood cabinets can be bought at a very good price. Most painters can do a good job of finishing these cabinets either before or after installation. Check their experience before selecting this option. Painted wood cabinets can be finished at the factory or finished on the job. If color in the kitchen is important, the latter may be your best choice.

Stained wood cabinets are finished at the shop or factory in an environment that ensures an excellent application and curing of the finish. There are many choices of stains.

Plastic laminate cabinets are made of wood or wood products and then covered with a plastic laminate. In some cases, the laminate is cemented to the inside and outside of the cabinet. In other instances only the door (all sides) and the front of the cabinet are covered with the interior being painted. The plastic laminate comes in a wide variety of colors and patterns and is easily cleaned.

One option available for paneled door cabinets consists of panels that are removable and reversible with different finishes on each side.

Soffits or No Soffits?

One question that often arises is whether to select kitchen cabinets that run all the way up to the ceiling. That top foot of the normal 8′ ceiling usually requires some sort of stool or ladder to reach, but it is additional storage space that comes in handy for seldom-used items.

If you have selected factory-made cabinets, you may not have any choice in wall cabinet height. Most such cabinets are made only at the size that requires a 1′ soffit down from the ceiling or it may be left open (see Figure 18.1).

The decision with regard to the soffit for kitchen cabinets must be made before the completion of the framing so that the carpenters can install the 2×4s needed to hang the drywall and secure the wall cabinets if soffits are required.

Bath Vanities

Bathroom vanities are usually supplied by the same source as the kitchen cabinets. The finish, material and color are a matter of personal choice and need not be the same as those for the kitchen cabinets.

The normal height of an installed vanity with top is about 30″ to 32″. This is low for a tall person. Consider ordering your vanities so that the height fits the user. If your vanities are factory built with only one choice of height, the trim carpenter can build a platform on which to mount the vanity to give the desired height.

Cabinet Countertops

Plastic laminate tops are available in two choices: custom and postformed (see Figure 18.2).

FIGURE 18.1 Wall-Hung Kitchen Cabinets

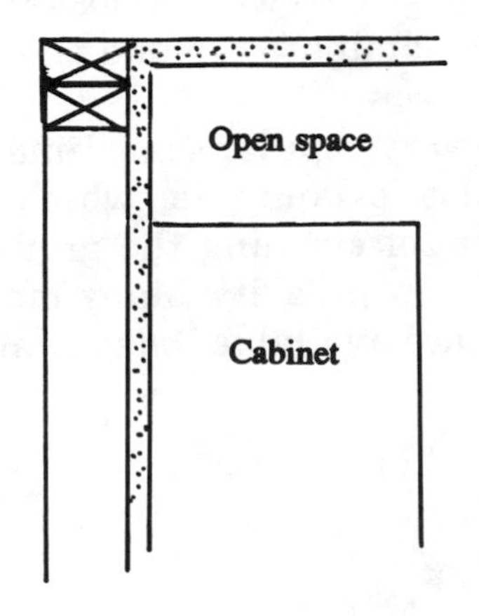

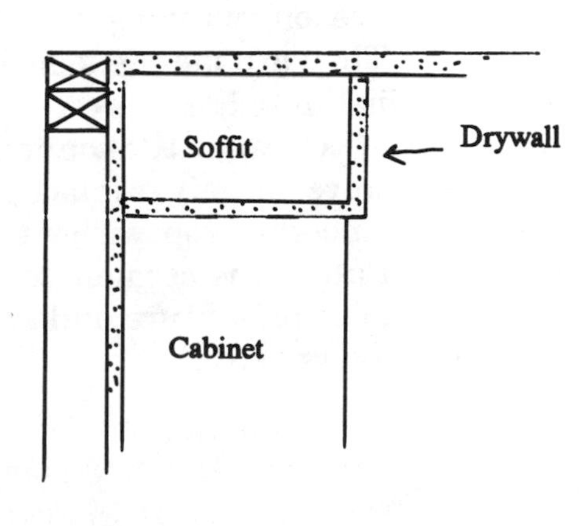

FIGURE 18.2 Countertops

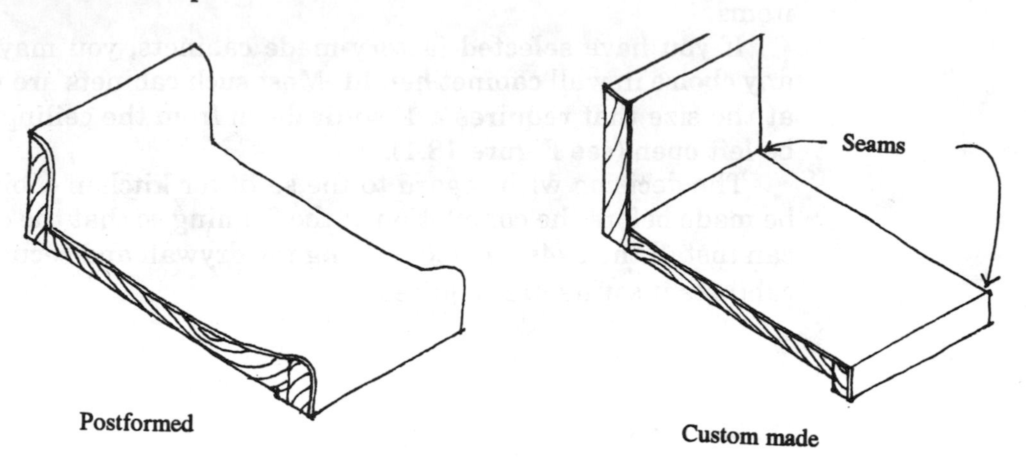

The custom-made top has the advantage of greater flexibility in shape and form. It has a seam located between the backsplash and the top, which is difficult to keep clean. Another seam in front is exposed to wear and, like all laminate seams, may part when the cement fails. If it is properly cared for, however, a well-made top should last 20 or more years.

The postformed top is molded under high pressure by machinery especially designed for this process. The result is a rounded back joint with no seam that is easy to clean and a lip on the front, also without a seam, that will prevent liquid spills from falling on the floor. The laminate used in the manufacture of this top is thinner than that used for the custom top so that it will bend and adhere properly to the base.

Countertops made of solid surfacing such as Corian, Nevamar, Surrell, Fountain Head and Avonite, with or without integrated sinks, are excellent choices for kitchens and bathroom vanities. They are long wearing, easy to install and go well with almost any cabinet. They are more expensive than the laminate top and choice of color is limited.

Ceramic tile countertops were very popular some time ago. They were largely replaced by the plastic laminate, which offered a smoother top without the problem of cleaning the grouted joints. Lately, the ceramic top is regaining popularity. Many more choices of color, texture and shape of tile are available today than 20 or 30 years ago.

Other Cabinet Work

If your house plans contain other cabinet work, such as bookcases, you can have this work done by the same source as your kitchen cabinets and bathroom vanities, or you can have it done on the job by the trim carpenters and finished by the painter. Your choice should depend on the complexity of this additional cabinet work. Most trim carpenters are perfectly capable of building good-looking bookcases, and most painters are capable of a good finishing job. Question the trim carpenters about what type of wood they prefer to work with. Inquire about their experience and ask to see some of their work. Generally, this approach to cabinets is less expensive than using custom-made or factory cabinets.

Costing

If your selection of cabinets is through a custom-cabinet maker or a cabinet store, either should furnish a complete bid including cabinets, tops and the labor to install them.

If you are buying your cabinets from a building supply store or some other source that does not provide installation, the cost of the materials is provided by the supplier and the labor cost from an installer such as the trim crew.

Management

- Order your cabinets early. Some suppliers take two months or more to deliver them. When you decide on your cabinet source, ask how long it will take and then add another month. Schedule your order based on this figure.
- Cabinets with countertops must be installed before the plumbing, electrical and trim work can be completed.
- Vinyl floor covering should be installed after the kitchen cabinets and the bathroom vanities are in place. See Chapter 19 for further details.
- One of the most prevalent problems with cabinet work is warpage of the doors. To lessen the chances of this failure occurring, avoid selecting cabinets whose doors are made with thin plywood (¼″ or less) let into a ¾″ wood frame.

19

Flooring

Wood

The most popular wood flooring choices are white and red oak. These are hard woods with excellent wearing qualities. The standard type comes in widths of 2¼″ and random length. Wood flooring is normally applied over the plywood subfloor and nailed to the floor joists.

To achieve a more custom look, oak flooring can also be supplied in random widths with or without a distinct V–groove between the planks (see Figure 19.1). A very colonial look can be obtained by driving two or three exposed, large-headed nails at the ends of the planks.

FIGURE 19.1 Random-Width Wood Flooring with V–Groove and Boat Nails

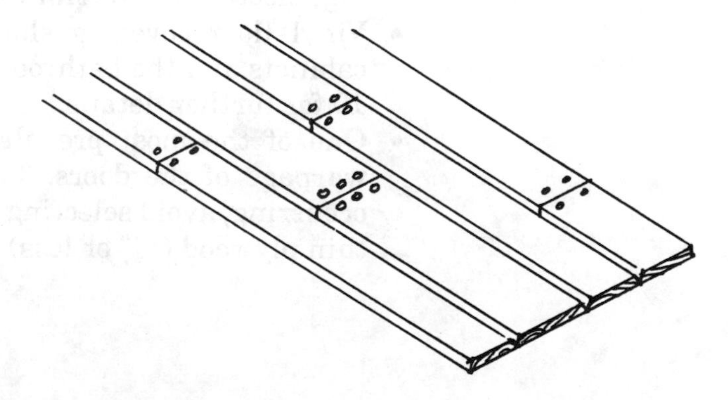

FIGURE 19.2 Parquet

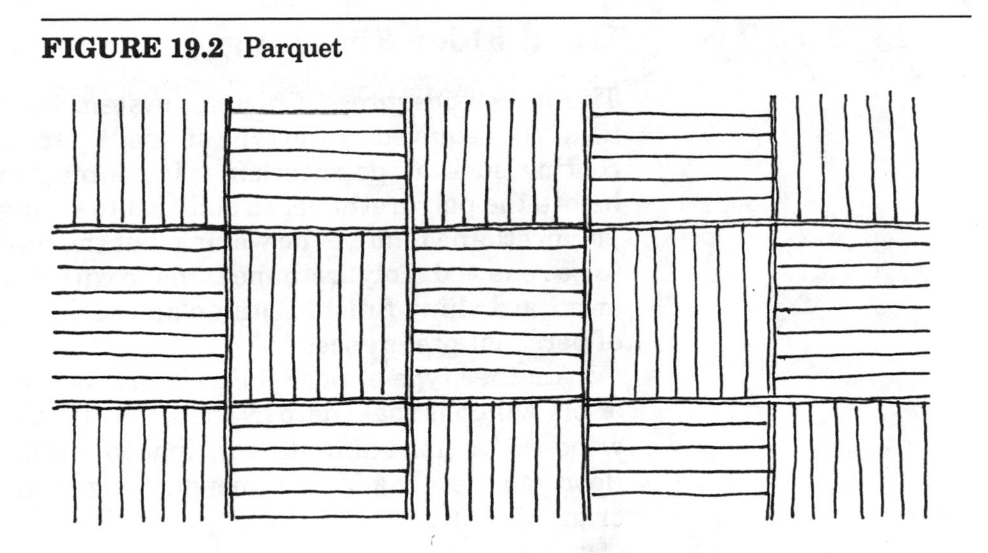

One of the best nails to use for this purpose is the large-headed, heavily galvanized nail used by boat yards. When it is applied to the flooring, this type of nail gives the appearance of being old and hand made. If the nail is driven flush with the top of the flooring, the soft galvanizing permits the use of sanding machines to finish the floor without ripping the sandpaper. In addition to the exposed nailing at the end of each plank, the usual blind nailing along the tongue side into each floor joist should be applied.

Pine can also be used for flooring material in standard or random width. It is a softer wood than oak and will show wear before the oak. It has a beautiful grain and takes staining well.

Parquet flooring consists of small pieces of wood combined into a square tile (see Figure 19.2). Some parquet flooring is very thin and requires an underlayment of particle board over the subfloor. Parquet is more difficult to finish since all sanding must be done with fine paper in order not to make large scratches across the grain.

Prefinished Wood Flooring

There are many different kinds and styles of prefinished wood flooring from plank to parquet. The finishes are excellent and generally long lasting. Follow the manufacturer's instructions for installing this type of flooring.

Wood Floor Finishing

The most widespread finishing system for wood floors consists of filling (if required by the type of wood), sanding and applying a hard coating such as polyurethane. If staining is desired, it is applied before the polyurethane. In addition to giving floors a pleasing look, polyurethane requires no waxing, but the finish will eventually have to be renewed. Polyurethane is made with either a matte (soft shine) or a hard shiny finish. Both do an excellent job; your choice is one of personal preference.

Another type of floor finish is the commercial penetrating resin or oil, which is soaked up by the wood. This finish does not cover the wood with a hard film. Its principal advantage is that to refinish the floor you merely add more penetrating resin without removing the original coating.

Carpeting

Carpet construction Virtually all residential carpets are tufted, that is, yarn loops are inserted into a backing material forming the carpet pile. If the carpet is to be cut pile (the most popular type) the tips of the loops are sheared off for a smooth finish. If the carpet is to be loop pile, the loops are not sheared.

Quality and performance The carpet should be warranted against stain and dirt resistance as well as protection against static and wear. A dense, thick pile wears longer, resists crushing and matting better and retains texture longer in heavy use. Compare density by bending two carpet samples as they would be bent over a stair tread. The higher-density carpet will show more pile fiber and less backing material. In carpets of equal density, the one with the higher or the heavier pile will perform best. *In general, the deeper and denser the pile, the better the carpet will perform.*

Carpet cushion Properly installed, this item will add 17% to 50% to the carpet's useful life. A medium thick pad is the best place to start. Thin pads can tear, wear or disintegrate too quickly. Ultra thick pads can be too soft for comfortable walking and balance.

Carpet Styles

- *Velvets,* the most elegant choice, provide an ultra-smooth sweep of rich color.

- *Saxsonies* or *plushes* are not as formal as velvet but are easier to live with.
- *Textured saxonies* are the most casual and rugged of pile carpets.
- *Level loops* have loops that are easily visible; this style blends with any decor.
- *Multilevel loops* display a carved pattern for a more random effect.
- *Cut and loop styles* or *traceries* consist of cut pile and loop pile combined. This style offers an infinite variety of effects.

Color Selection

- Light color carpets will make the room seem larger, particularly if the walls are white or a light tint of the carpet color.
- In cold, snowy climates red, yellow and brown carpets can warm up north-facing rooms.
- In warm climates, blue, green or violet carpets can be used to "cool off" south-facing rooms.
- Warm reds and oranges create an active atmosphere, which is great for family rooms.
- Cooler blues and greens generate a tranquil setting for bedrooms and other quiet areas.

Installation

Carpeting should be installed over underlayment nailed to the subfloor with at least one layer of building paper in between. If the room has little moisture, the underlayment can be ⅝″ thick particle board, an inexpensive material. If the carpeting is installed in a bathroom or kitchen, however, the underlayment should be ⅝″ thick plywood. This material is more expensive than particle board but will usually not be affected by normal amounts of moisture found in kitchens and baths. Particle board, on the other hand, tends to absorb moisture, swell and disintegrate.

If carpet is to be installed over a concrete floor, first lay down a cover of high-density fiberboard (usually ½″ thick), followed by the carpet padding, then the carpet. This system will provide a warmer, softer floor with better personal comfort.

Vinyl Tile

Vinyl tile and sheet material can make a fine floor covering for the bathrooms, kitchen, sun room or any other room, for that matter. The choices in pattern, color and quality are many.

This material should be installed over plywood underlayment. Most manufacturers will void their warranty if it is laid over particle board.

Ceramic Tile, Earthstone, Slate and Stone

Ceramic tile, earthstone, slate and stone all make excellent flooring. They will show practically no wear and require little maintenance. For the best results, they should be laid in cement mortar and, with the exception of bathroom floors, they should be laid over a dropped subfloor. See Chapter 5, "Framing."

Slate and stone can be very heavy, depending on size and thickness selected. For these heavier materials, the floor joists in the area should be installed 12″ on centers to give more strength.

Brick

Brick makes an excellent floor and can be laid over concrete or a wood subfloor with mortared or unmortared joints. Use the special brick made for this purpose. Remember that brick is a heavy material, so install the floor joists 12″ on centers in that area.

Most brick is porous and should be sealed. Use a phenolic type of concrete sealer and apply at least two coats. If the brick has been installed with mortared joints, be sure that the material is completely dry before applying the sealer. Drying time can be as long as eight weeks.

Costing

In the bidding process for wood flooring, it is best to ask the contractor to provide all material and labor. Although many building supply houses sell wood flooring, you will simplify the management task if you have the contractor supply all material. If you feel that you can get the job done at less cost by buying your own flooring, then ask for bids with two prices, one with the contractor supplying all materials and labor and the other with the owner providing the flooring and the contractor all other material and labor. Then select the method that is more advantageous to you.

The entire cost of carpeting is provided by the contractor in the bid. It is best if the bid is based on actual measurements from the blueprint rather than a statement of the required number of square yards of area to be covered.

The entire cost of the vinyl floor covering should also be provided in the bid from the contractor, who will need an exact drawing of the area, particularly if the specifications call for sheet material rather than square tiles. Most vinyl sheet material is manufactured in rolls 6′ and 12′ wide. The bidding contractors will select the widths that best fit your plan.

Unlike carpeting, vinyl sheet material cannot be cut and patched without creating undesirable seams and pattern mismatches. In most cases, there will be waste that the owner must expect to pay for. For example, in a room that is 12′×16′, a piece of vinyl 12′×16′ will do the job nicely with practically no waste. In a room 10′ wide and 16′ long, the contractor must still supply a piece of sheet vinyl 12′×16′ and charge the owner for the full amount although the room is not that big.

The cost of all labor and material for ceramic tile, slate, earthstone, stone and brick should be included in the contractor's bid.

For your ceramic tile floors, specify whether you want tile or wood baseboards. Either can be used. It is a matter of personal choice, the tile having the advantage of low maintenance. The wood base gives a more traditional touch and, since painting or staining is required, offers the option of changing the color from time to time. The ceramic tile contractor will furnish all material and labor but will expect you to furnish and have installed any wood base.

Management

- Unfinished wood flooring can be installed before or after the drywall or plaster. Trim work, plumbing, heating and cabinet work cannot be completed until the wood flooring has been installed.
- Protect prefinished flooring immediately after it has been laid. Heavy canvas does a good job.
- Wood floors should be finished only after all other interior work has been completed (with the exception of laying carpeting). During the finishing process and for three or four days thereafter, no one except the floor finishers should be permitted inside the house. Activity stirs up dust and other particles in the air, which may settle in the wet finish and ruin it.

- After the floor finish has completely dried, protect it by laying flint paper or some similar material on those areas where people will walk.
- The laying of the carpeting is usually the last job to be done. It is the owner/general contractor's responsibility to have the underlayment down (by the carpentry crew), before the carpeting is installed.
- In all areas in which a wood baseboard with wood shoe molding is part of the trim, it is best to lay the flooring first, then apply the trim. One exception to this rule is carpeting. The carpet is laid after the baseboard is installed and the carpet layer draws it up to the baseboard using special equipment. Shoe molding is generally not used with carpet.

  If baseboard heating units such as baseboard electric heat or baseboard radiators for hot water heating systems are in rooms that will be carpeted, the radiator element should be raised higher on the wall so that the carpet material does not interfere with the free flow of air up under the element and out through the top.
- Floor underlayment and wood flooring (unfinished) should be installed before the base cabinets.

20

Painting and Decorating

Painting and decorating produce results that you will be looking at for many years: It is most important that the results be good ones. Pick quality materials and the best labor. The paint will last longer and look better throughout its life.

To ensure that high-quality materials are used, visit one of your local paint suppliers and pick out the top of the line for each painting task. Use this paint designation in your specifications, indicating that either this paint or its equal must be used. The brand of paint need not be a national one, since many excellent paints are made at local factories, but if you do use a local paint, be sure that it is fine quality. Ask the seller to give you the address of a local house painted with his paint, then take a look at it. The major cost of the total paint job is the labor. It just doesn't make any sense to pay for good labor and use a mediocre paint.

Exterior Painting

Your choices of paint are:

- *Alkyd oil-based paint* has been the standard house paint for many years. It is now being strongly challenged by the latex water-based paint. Alkyd will do an excellent job on wood and metal and offers unlimited choices of color. Because it has a hard finish, alkyd should be used on trim and porches.

- *Acrylic latex* (vinyl or acrylic) water-based paint is available in an unlimited range of colors and can be custom-mixed by most dealers.
- *Stain* is available in both water-based and oil-based transparent colors. Exterior stain is available in semitransparent oil-based or acrylic latex solid formulas. Semitransparent stain can also be used on the interior. Solid color stains on wood shingles are difficult to remove.

In making your selection, consider the following points:

- Acrylic latex paint is less costly than alkyd oil based paint, lasts almost as long, is easier to apply and clean up and its labor costs should be lower. However, alkyd dries harder and is more scrubbable and so should be used for doors, window frames, shelves and other wood in heavy wear areas.
- Both alkyd oil and acrylic latex paint require a primer over bare wood. An oil-based primer is best, although an acrylic latex primer is a good second choice. Oil bases should be applied only over dry surfaces, but latex can be used on dry or damp, not wet, surfaces. If you use an oil-based primer, it can be finished with a coat of either alkyd or latex paint.
- If you live in a mildew prone area, antimildew additives can be put in both acrylic latex and alkyd oil paint. Most quality paints already have some mildew fighter already mixed in. Latex is more resistant to mildew than alkyd, but alkyd is more scrubbable.
- A three-coat system (primer plus two additional coats) will last almost twice as long as a two-coat system (primer and one additional coat).
- From the standpoint of cost alone, the three-coat system is the logical choice the first time the material is painted.
- If poor-quality paint is used on areas above brick or other nonpainted material, chalking of the paint down over the brick may occur and spoil its appearance. If you select a top-quality nonchalking paint, this problem should not occur. There are some quality paints purposely formulated to scale off some of the top layer so that the remaining paint maintains its bright color and stays clean. These paints should also not be used above unpainted brick and similar surfaces.
- If exposure of the wood grain is important, use a semitransparent oil-based stain. The number of coats of the stain depends on the desired color.
- Exterior stains do not last as long as paint and should never be clear coated, since the sun will erode the clear material, which cannot block the sun's ultraviolet rays. Stains do not peel and crack as paint does.

Back Priming

All bare wood exposed to the weather should be back primed to reduce warping and cupping of the wood and to prolong its life. The painter should apply on the *nonexposed side of the wood* one coat of same primer to be used on the exposed side. Don't forget the exposed ends of butt joints in the siding and trim.

Caulking

Caulking, a job normally done by the painting contractor, is required regardless of which paint or stain is used. Caulking is applied at all joints in the exterior painted wood surface to seal around the trim of windows and doors, the joints where the lap siding abuts the corner boards and so forth. Latex caulking is one of the best choices for house painting considering cost, durability and ease of application. Although caulking with silicone added lasts longer, it is more expensive than the latex version. New caulking compounds can stretch with house expansion and remain in place. This reduces the need to replace the caulking every few years.

Mildew

Mildew, a living fungus, is a product of the environment. It will normally flourish in warm, humid, shady areas, which encompasses a large part of the country. No paint or stain can be made completely mildewproof, but additives can retard the development of mildew. For this reason, it is important to prime and paint the exterior of the house as soon as possible so that mildew cannot form on the bare wood or between the coats of paint or stain.

Interior Painting

Select a flat or satin finish paint for walls and ceilings, except in bathrooms and kitchens, where a semigloss or gloss enamel paint should be used for easier cleaning and moisture resistance. The enamel paint also gives better protection to the drywall or plaster.

For the interior penetrating stains, only one coat is usually necessary. Additional coats can be applied, however, to get the desired depth of color. Apply a clear sealer over the last coat for easier cleaning.

Bare wood trim should be painted with an enamel paint or stained. Acrylic latex is a good second choice if the wood is primed with an oil based primer.

In selecting colors for the interior consider the following:

- Dark colors make a room seem smaller but give a more dramatic appearance.
- To create a more restful room, use pastels.
- A white or light colored ceiling gives the impression of height, whereas a dark color appears to lower the ceiling.

Wallpaper and Wallcovering

If you select wallpaper or wallcovering for the final finish on some of your walls, here are some tips:

- For bathrooms or kitchens, choose a vinyl or vinyl-coated wallcovering that can be scrubbed.
- Behind the stove use a fabric-backed-vinyl paper. This is the toughest kind.
- Using strippable wallpaper will facilitate its removal if you redecorate in the future.
- If you want to make the room seem taller, select a vertical stripe for the wallpaper pattern.
- If you want to make the room seem larger, select a pattern with small figures and light colors with large amounts of clear space.
- Some wallpapers are water or moisture resistant. If you select one that is not, however, you can use a wallpaper sealer after it has been installed and the paste has dried. Sealers can change the color of the wallpaper, so test a sample before covering the entire wall.
- Leftover wallpaper can be cut to the proper size and used to cover switch plates or to line drawers and cabinet shelves.
- For walls that are untreated drywall, plaster or plaster veneer, sizing should be applied before hanging the wallcovering. Without this sizing the paper may not adhere properly to the wall. In addition, repapering in the future will be difficult because the wallcovering is hard to remove and the drywall or plaster may be damaged.

Special Ceiling Finishes

There are two popular methods for finishing ceilings (or walls, too). One is for the drywall finisher to cover the entire ceiling with a coat of texturing paint or joint compound and then go over it with a broom or similar device, putting a swirling pattern into the still-wet compound. The other method is for the painter to spray the ceiling with heavily textured paint material.

Cost

The cost of the painting, both labor and material, should be included in the bid of the painting contractor.

The cost of the wallpaper and mounting material is provided by the supplier. Have the supplier compute the amount of paper needed so you are assured of not running short because you forgot to allow for pattern matching and other waste. The labor for installation can be provided by the supplier or it may be part of the painter's bid. Many painters hang wallpaper.

Management

- Schedule the painter to back prime all exterior wood before it is applied to the framing.
- Apply the paint to the exterior of the house as soon as the job can be done, particularly if you live in a mildew-prone area. Primers must be finish coated within a week of their application or they begin to chalk and erode. Paint will not stick to chalking surfaces.
- Make certain that the top and bottom edges of all doors, particularly the exterior doors, are painted as fully as other areas. Most door manufacturers will void their warranty if this is not done.
- If your windows require painting, check to see that the seam between the glass pane and the wood frame has been sealed with paint. There should be a small strip of paint actually on the glass. The same procedure applies to wood doors with glass inserts.
- Your paint contractor may want to spray the interior walls and ceilings with the prime coat before the wood trim is installed and after the drywall has been finished and dried. This is a normal and efficient practice that should save on labor costs.
- Complete the interior painting or staining of a room before hanging the wallpaper or wallcovering.

- Before accepting the color selection for exterior paint or stain, have the painter show you samples on scraps of the exterior wood material. For interior walls, use samples on a small section of the wall or on pieces of drywall scrap. Since paint colors will vary with the type of lightings, view the sample colors under the type of lighting you plan to use in each area.
- If you select wallpaper or wallcovering, be sure to order more than you will need for the initial hanging. This will give you some extra to make repairs to the wallpaper later should the need arise. Wallpaper designs are often taken out of production without notice and are not available later.

21

The Finishing Touches

Landscaping

Landscaping serves two purposes: it drains water away from the foundation and it beautifies the lot. As a minimum, landscaping should include the final grading so that the drainage is accomplished and then the grade should be stabilized by planting grass or other ground cover, which will give the dirt stability from water and wind erosion. Usually, a layer of straw spread over the area will minimize erosion until the ground cover has grown enough to do the job. If the grading and grass planting is performed by a contractor, there should be a clear understanding concerning responsibility for repairs of damage due to washout.

The completion of the landscaping is a matter of personal choice. You may elect to have the job done by a professional nursery, most of which will provide you with a recommended plan. This is the most expensive option, but if you need professional help, it is usually worth the additional cost. Most nurseries will guarantee the survival of their plants.

You may also decide to do your own planting by buying plants and providing the labor. Under these circumstances, however, the plant suppliers will probably not guarantee their products.

Garage Doors

Garage doors are a complex part of your house requiring a special skill for proper installation. For this reason, you should have a contractor who specializes in this work provide both the labor and the material. If your plans include radio-controlled doors, this approach becomes even more important.

The overhead door is the most popular, primarily because it is readily adaptable to radio-controlled electric door openers. If your plans include two single (about 8′ wide) doors, an electronic door control system with two frequencies is preferable so that each door can be controlled separately.

Decks

Wood decks should be built with lumber that can withstand the rigors of constant exposure to the weather. Redwood, cypress and cedar are all excellent selections, but they are expensive. Salt-treated southern yellow pine is also an excellent lumber for decks and it is less expensive than the other options. Be sure to order grades of salt treatment labeled .25CCA for lumber to be used above finished ground level and .40CCA for material to be used in contact with the ground.

Southern yellow pine has a greater tendency than most other lumber to warp when exposed to the sun. To compensate for this tendency, design your horizontal railings as indicated in Figure 21.1. Nailing the two pieces perpendicular as shown will reduce warping. In addition, warping will be reduced if the railing spans are less than

FIGURE 21.1 Deck Construction

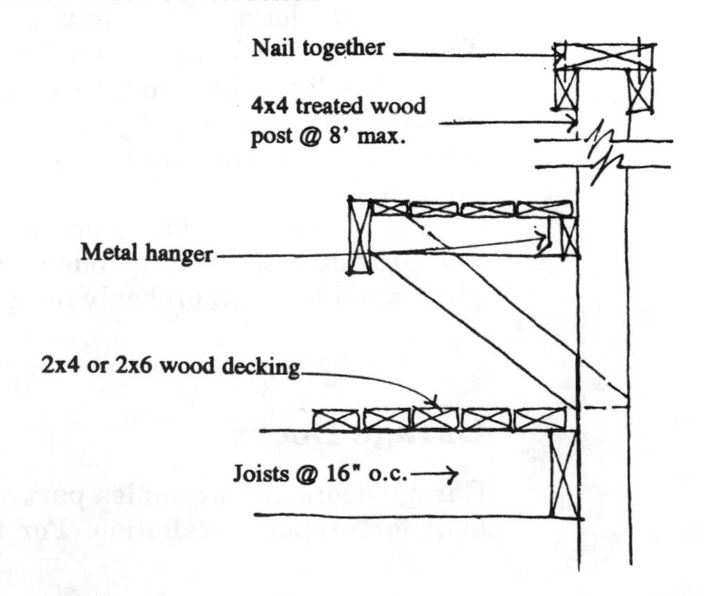

8′. Benches add to the stability of the structure and improve the usefulness of the deck.

Use *hot dipped* galvanized nails or aluminum nails for the deck construction. Other types will begin rusting soon after the structure has been completed.

Metal Railing

If your plans include metal railings on the front steps or elsewhere, you can buy the railing already built and have it installed, or you can contract with a firm specializing in this work to design and install the railing. These railings are usually made of steel or aluminum.

Check your local building code for the railing requirement. Even though your plans do not include railings, the code may require them, especially if the grade of your lot has created higher outdoor steps than the original plan indicated. This situation often occurs if the plans were not specifically designed for your lot.

The main problem with metal railing is proper installation. Two methods are illustrated in Figure 21.2. The best method installs the vertical pieces of railing down into the brick or concrete in a large hole, which is filled with molten lead. This method gives the railing

FIGURE 21.2 Railing Installation

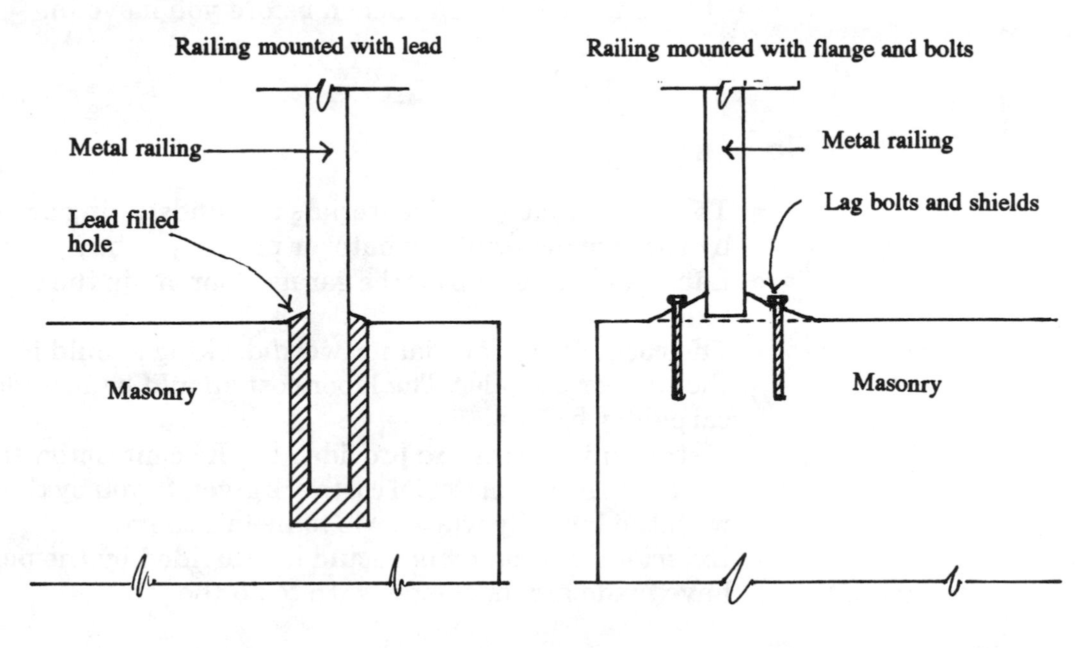

the strength to withstand the pushing and pulling it will have to undergo in normal use.

A second method is to install the railing with bolts made especially made for masonry and concrete (there are several varieties). They should extend into the masonry at least 2½″.

Clean-up

At various points throughout the construction period, you will have to do a major clean-up of the area. You should plan for at least three. The first will be needed after the framing, exterior trim and siding have been completed and the shingles installed. The second will be after the drywall or plaster has been finished. And the last is when the house is basically completed but before the final grading is done.

Many of the wood scraps will be useful to you in the future for fireplace kindling or woodworking. If you can find a storage space that does not interfere with the progress of the construction, save them. *However, do not use scraps of treated lumber in your fireplace or stove.* The chromate copper arsenate (CCA) preservative emits poisonous fumes when burned, possibly causing cramps, headaches, earaches, diarrhea, rashes and other symptoms.

To get these clean-up jobs done, find someone with a truck who can do the work and haul the debris to the nearest dump. Remember to avoid paying wages and getting involved in a lot of extra paper work: have a contract for a fixed price for each clean-up.

Crews can also be found to clean the inside of the house to include bathrooms, kitchen, paint scraping of windows and other cleaning chores that should be accomplished before you move in.

Costing

- The costs of labor and materials for landscaping are provided by the contractor plus whatever plants you buy yourself.
- Labor and materials for the garage door are in the contractor's bid.
- The cost of the material for wood decking should be given by the lumber supplier. The labor cost should be included in the carpentry bid.
- Metal railing costs are provided by the contractor. If you buy the railings, the material costs are given to you by the supplier with the labor by whoever is to install them.
- Exterior clean-up costs should be provided by the person you have made arrangements with to do the job.

- Interior clean-up costs are provided by the contractor.

Management

- Final grading and landscaping may reveal construction debris such as pieces of masonry and wood that should be removed from the lot. Make sure that all loose wood has been removed from the crawl space, particularly if you are located in a termite prone area.
- Contact the garage door installer to determine what framing features should be completed before the door is installed. Have this job done by the carpenters before they leave.
- Have the garage door installed early to help secure the structure and to give you protected space for temporary storage of other building materials such as interior trim and cabinets. The garage floor should be poured before the door is installed.
- Before the floor is poured, check the forms to ensure that the floor is exactly level at the line where the bottom of the garage door will rest so there will be a good seal at this point.
- Final grading should be completed before the deck is installed, particularly if the deck is large. If this is not done, it will force the landscaper to use hand methods under the deck and thus increase your costs.

22

House Expansion

This chapter discusses construction techniques related to the expansion of an existing house rather than building an all-new house. The reader is assumed to be familiar with all of the other chapters and the appendixes.

Before going ahead with plans for expansion:

- Determine whether or not your lot can accommodate the size of the expansion and remain within the setback lines established by the local government.
- If your present plumbing system uses a septic tank and drain field and your expansion includes additional plumbing that will increase sewage, have a professional contractor check your system to make sure it can handle the additional load. In some cases, a larger tank and/or drain field may be necessary requiring additional space on the lot.
- Make sure your plans have been approved by local architectural review boards if required.

In many cases of house expansion, it is more difficult to arrive at a practical, well-done plan than in a completely new construction. For this reason, during the planning phase, it is usually wise and less expensive in the long run to get professional help, such as an architect, at least on an hourly consultant basis.

Compared with construction of a completely new house, in many cases it is difficult for the contractor to determine exact costs in house expansion. For this reason, many contractors will not offer a fixed price for the job.

Expansion Choices

Additional room can be added to a house in numerous ways, several of which are described below:

Expand Right or Left

This type of expansion prolongs the existing roof (see Figure 22.1). "A" is a simple solution, but the overall look of the house will be more attractive if the wing or wings are narrower than the main house so that the roof line is different, as in "B." In addition, "B" is a better choice since it minimizes the problem of matching the shingles between the old and new sections and will make any differences in shingles less conspicuous.

Expand to the Rear

You can add a wing that is perpendicular to the main line of the existing house (see Figure 22.2).

Convert the Existing Garage

This option calls for converting the existing garage to living space and adding a new garage. Most houses are built with the living area over crawl space (2′ to 3′ above the lot level) with the concrete garage floor poured at the level of the lot. If this difference can be spanned by one or two steps (about 8″ each), this choice has merit. If the difference is more than two or three steps, however, other expansion choices may be preferred.

FIGURE 22.1 Expanding Right or Left

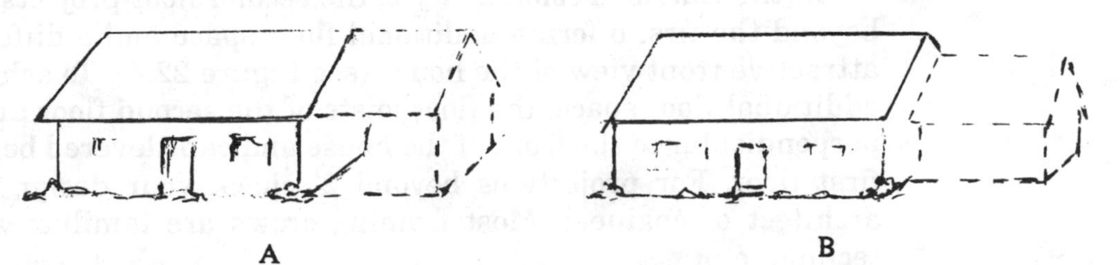

FIGURE 22.2 Expanding to the Rear

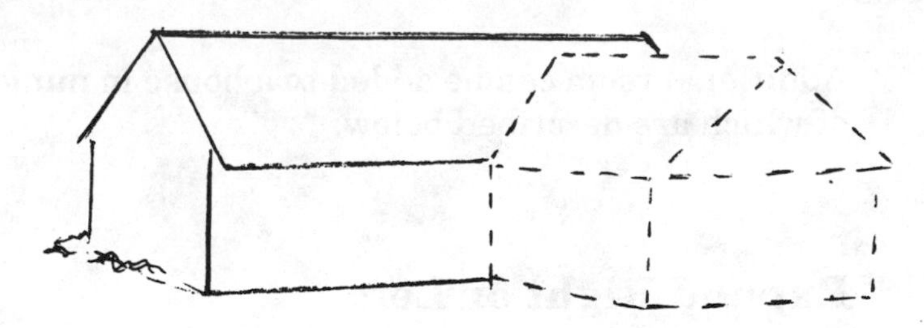

With this option, much of the construction already exists, and, with the exception of the garage-door area, the design features of the original house are retained.

Add a Second Story

Adding a full or partial second story can substantially enlarge an existing house. The footings and foundation should be checked by an engineer, architect or qualified contractor to see if they can safely carry the additional weight of a second floor. In addition, the ceiling joists of the first floor now become the floor joists for the second floor and may have to be replaced with stronger material, such as 2×10's or 2×12's to replace 2×8 ceiling joists.

Several styles of second floors are illustrated in Figure 22.3.

With the conventional colonial second floor, the first and second floors have equal floor space.

The salt box design uses the rear of the roof to provide both roofing and exterior walls for the rear part of the second story. Dormer windows can be used for additional light and ventilation.

The second floor of both the story and a half and the salt box have less floor space than the first floor.

The existing roof of the story and a half provides both the roof and the exterior walls of the second floor. Additional light and ventilation are gained through dormers in the front and rear, with shed dormers usually in the rear only.

In the Garrison colonial style, the second floor projects about 2′ beyond the first offering additional floor space and a different but attractive front view of the house (see Figure 22.4). To achieve this additional floor space, the floor joists of the second floor are placed perpendicular to the front of the house and cantilevered beyond the first floor. For projections beyond 2′ check your design with an architect of engineer. Most framing crews are familiar with this technique.

FIGURE 22.3 Two-Story Houses

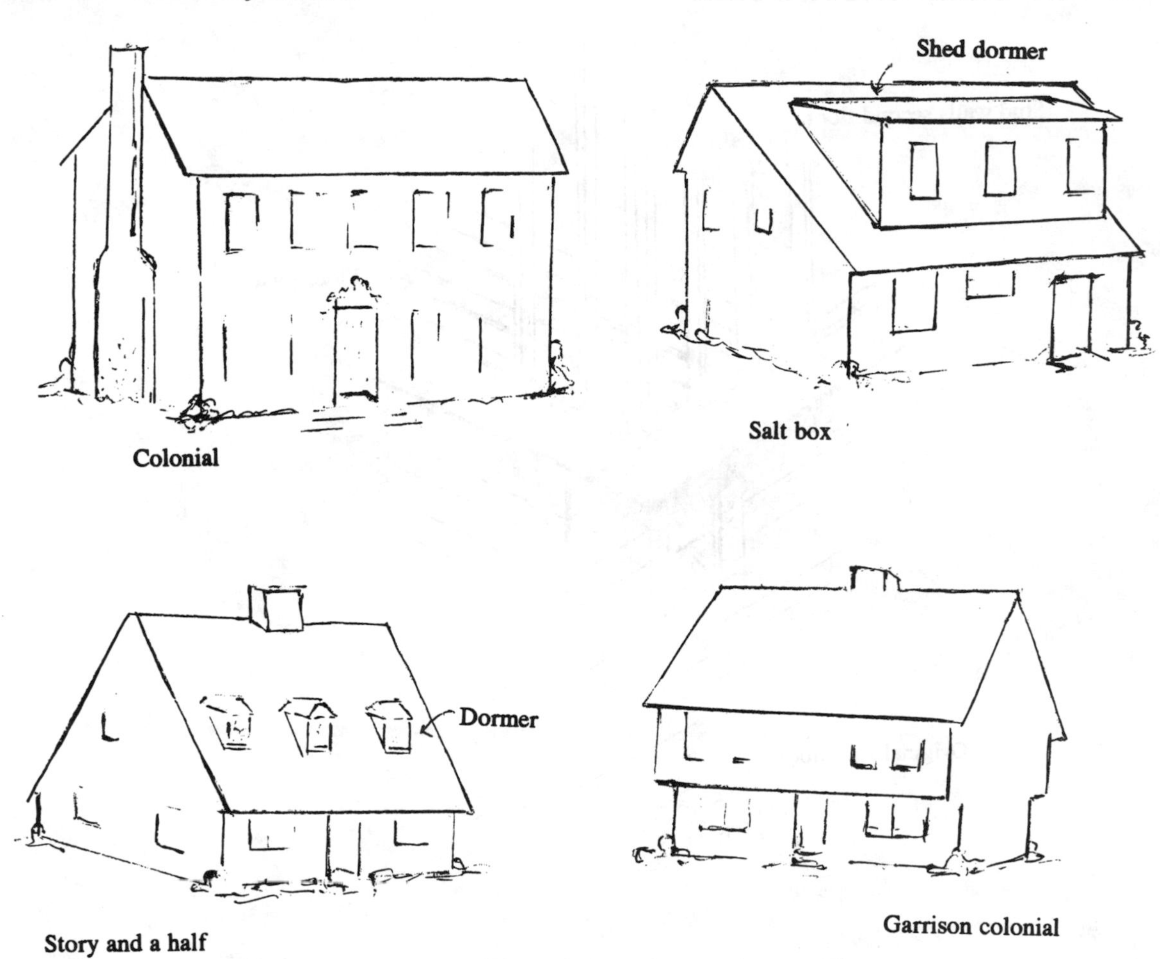

Making the choice of which type of two-story house to build is a matter of taste. Architects can be help you make this choice.

Use the Basement

If you have one, why not convert the basement to comfortable living space? The outer walls and floors are already built.

Is your basement damp? Several methods can be used for correcting this fault. First try to get rid of the dampness by applying masonry waterproofing material to the *interior* of the exterior walls. There are several choices on the market. After the job is complete,

FIGURE 22.4 Cantilevered Floor Joists

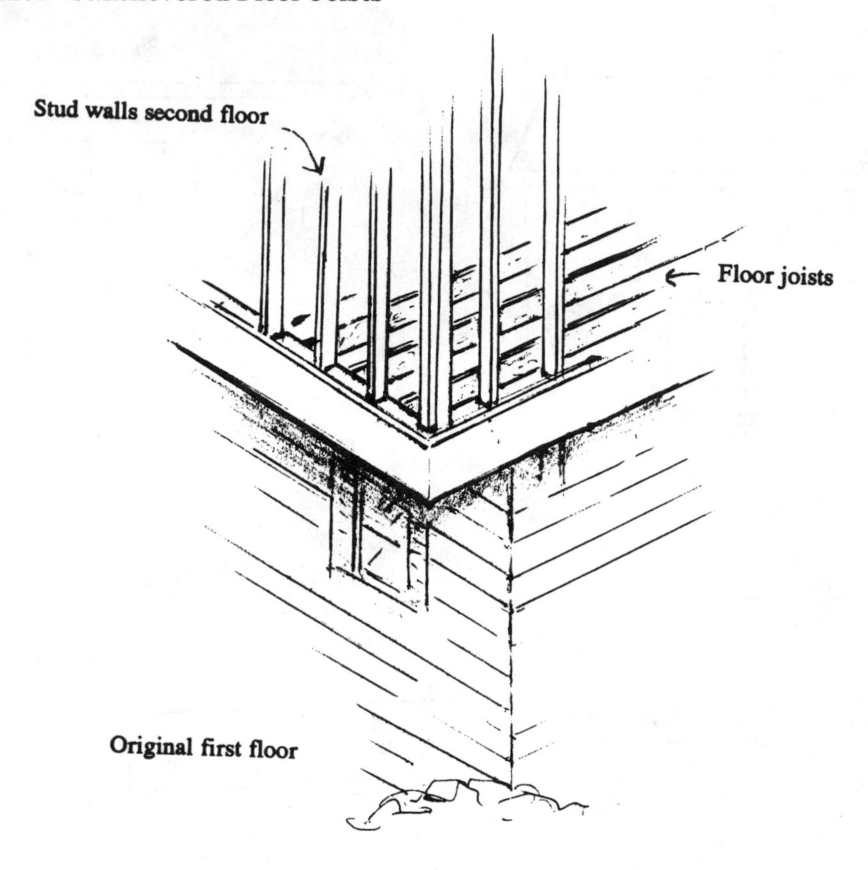

plug in an electric dehumidifier to remove all of the dampness and then wait about two weeks to see if the problem has been solved.

If the first solution does not do a complete job, try this method, which is more expensive but more effective. Cover the interior of the wall with rubberlike membrane (see pages 49 and 50), install a drain trough in the concrete floor next to the wall and drain this ditch into a sump equipped with sump pump. This work should be done by a professional waterproofing contractor. With this system, the rubber membrane will normally stop moisture from coming through the wall. Although moisture may come through the juncture of the wall and the floor, the ditch collects this moisture and moves it to the sump, where it is pumped out of the house.

The most effective solution is the proper preparation of the *exterior* of the basement walls. These walls, of course, must be exposed by excavation to permit the application of the rubberlike waterproofing membrane. In addition, a drain alongside the footings, as shown in Figure 5.4 (page 49) is installed. Again, this work should be done by a professional waterproofing contractor.

Basement living spaces offer an additional advantage over space above-ground in that the temperature of the ground in the winter is usually much warmer than the air above. The temperature of the ground is much less than outside air in the summer. These features produce an environment requiring significantly less energy to heat and cool.

Since concrete and concrete block provide little insulation, the finished basement will be much more comfortable by building a stud wall of 2×4's or 2×2's on the inside so that fiberglass or similar insulation can be installed between the studs adding effective insulation. These stud walls also provide space to run concealed electric cables, pipes and heating and cooling ducts.

The walls can be finished by applying drywall or paneling over the studs.

Like the garage modification, this choice also generally retains the original design of the house.

Enlarge a Detached Garage

The addition of a second story to an existing garage offers a great deal of privacy from the main house. Water, sewer and electrical power can usually be brought from the main house. Review the construction problems discussed in "Add a Second Story," on page 226.

An exterior stairway to the new second floor that avoids the garage is desirable.

Some expansion plans may distort the appearance of the original house to the extent that it is objectionable to neighbors and the review boards (subdivision, county or city) that must approve the plans. The primary legal basis for these objections is that the unattractiveness of the expanded plan may lower the market value of neighboring houses.

Note in Figure 22.5 how the addition of a second story on the garage of a one-story house results in an awkward design.

A second floor over a garage for a rancher is more practical if it can be built within the existing roof, perhaps with dormer windows.

Construction Techniques for Additions

Footings and Foundation

Review Chapter 4, "Footings" and Chapter 5, "Foundations." To minimize separations between the old and the new structure, the footings and foundation walls of each must be connected. If both the

FIGURE 22.5 Unattractive Expansion Plans

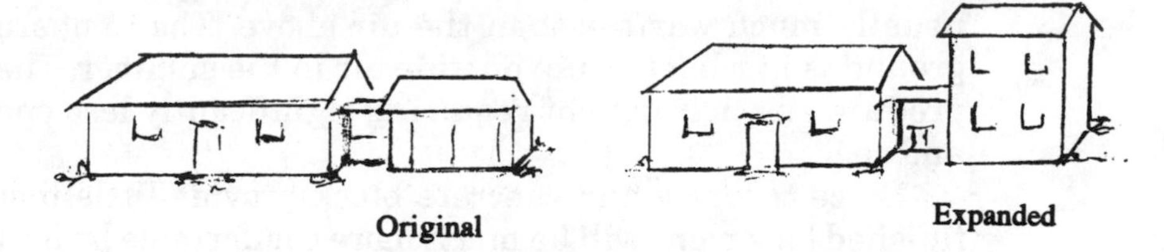

old and the new footings and foundations are concrete, steel reinforcing bars should be installed. Generally, two rebars of ½″ diameter are adequate for the footings with an additional two in foundations walls that are not over 3′ high. For a higher wall (if you are connecting an old and a new full basement walls, for example), include one additional ½″ rebar for each 2′ of additional height. When building the new concrete foundation wall, include a concrete pilaster as shown in Figure 22.6.

Install the rebar in the existing footings and foundation *before any new concrete is poured.* Drill the horizontal holes at least 1″ in diameter and 1′ deep for each rebar. The large hole diameter is

FIGURE 22.6 All-Concrete Footings and Foundation

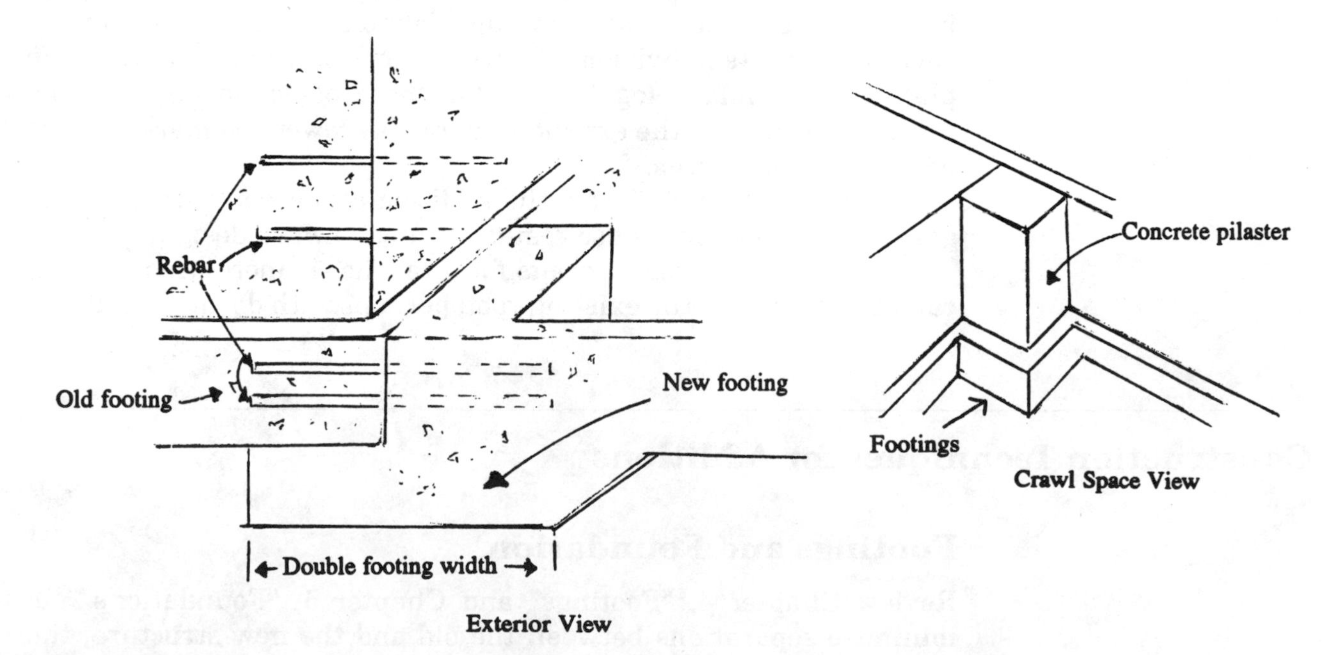

necessary to allow the newly poured concrete to move into the hole and bind the rebar to the old concrete.

In addition to the rebar, the existing footings should be undercut. The length of this undercut should be double the width of the footing, and the depth of the footing in the undercut should be at least 10″. When the new footings are poured, make sure that the new concrete completely fills this excavation to tie the old and new footings together.

If the existing foundation wall is concrete block, rather than using rebar to connect the new and the old, horizontal truss reinforcing (see page 46) can be used. In every other course that does not already have truss reinforcing, install an additional truss by cleaning out the mortar of the existing foundation wall back about 1′, then install the new truss reinforcing into this joint and run the truss completely around that course of block in the new foundation wall. Replace the mortar in the old wall. Install the concrete block pilaster with the new block wall for additional strength (see Figure 22.7).

Don't forget the vertical reinforcing in the new foundation wall if required.

FIGURE 22.7 Concrete Footings with Block Foundation

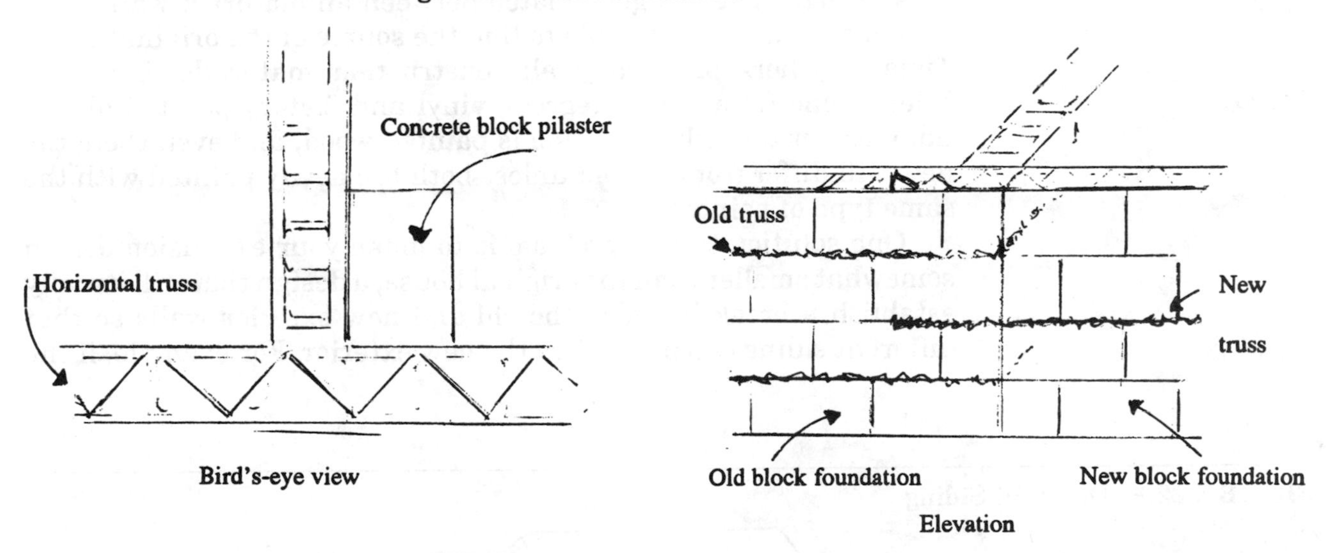

Siding on the Existing Wall

On that portion of the existing exterior wall where an addition is to be applied and this wall becomes an interior wall, should the masonry, wood, vinyl or aluminum siding be removed to expose the wood framing beneath? This is the owner's choice. If you want the new addition to have an interior wall of the same material as the siding, then go ahead without removing it. The new framing can be readily attached to the existing wall by using the proper bolts or nails.

On the other hand, if you want a plaster, sheetrock or a paneled interior wall, you can remove the old siding or masonry revealing the original stud wall to which the plaster, sheetrock or paneling can be attached. Another solution to providing a base for the plaster, sheetrock or paneling is to leave the existing siding and masonry in place and install a new stud wall in front of it on the new room side. In either case you have the space to run electric cables, pipes and heating duct work concealed behind the interior wall finish.

Matching the New Exterior Wall Cover with the Old

It is difficult to get a good match between an old brick wall and a new one even if you are able to find the source of the original brick. Time weathers practically all construction materials, including brick, stone (to a lesser degree), vinyl and factory-painted aluminum. About the only exception is painted wood, and even then, the new will differ from the old unless both are newly painted with the same type of paint.

One solution to this problem is to make your expansion design somewhat smaller than the original house, a design that will usually establish a break between the old and new exterior walls so that different siding can be used on the new exterior. For example, if the

FIGURE 22.8 Different Siding

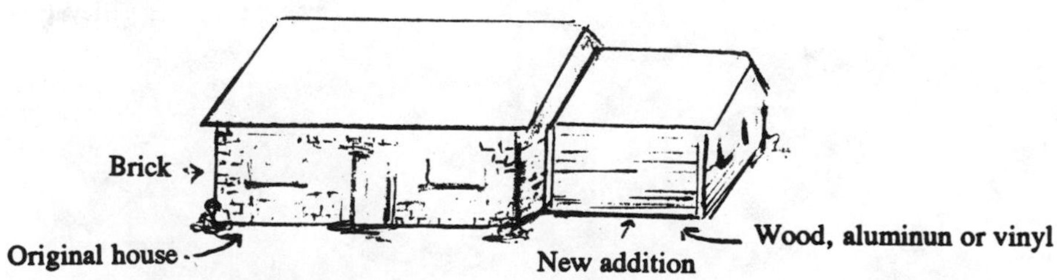

original house is brick or brick veneer, change to vinyl, aluminum or wood for the exterior of the new section (see Figure 22.8).

This design will also result in a break in the roof lines of the old and new sections of the house and deemphasize a difference in the color of the old and the new shingles.

Windows

Review Chapter 7, "Windows and Doors," to help you decide on the type of window you want in your addition and the kind of glass to be used. If your existing house has single-glazed windows, you should consider replacing them as part of your home improvement project.

If the design of your new section permits, place as much glass (windows and doors) facing south or near south to take advantage of passive solar heat. (See Chapter 13, "Solar Heating.")

Heating and Air Conditioning

Forced Air Systems

The duct system for the original house, (see Figure 22.9) was designed to move warm and/or air-conditioned air to the rooms of just that house. Extending that system to take on additional room(s) will badly unbalance the duct design and produce an unsatisfactory air flow for the entire house, both old and new sections, even if the furnace or heat pump can take on the additional load.

FIGURE 22.9 Typical Supply Duct System

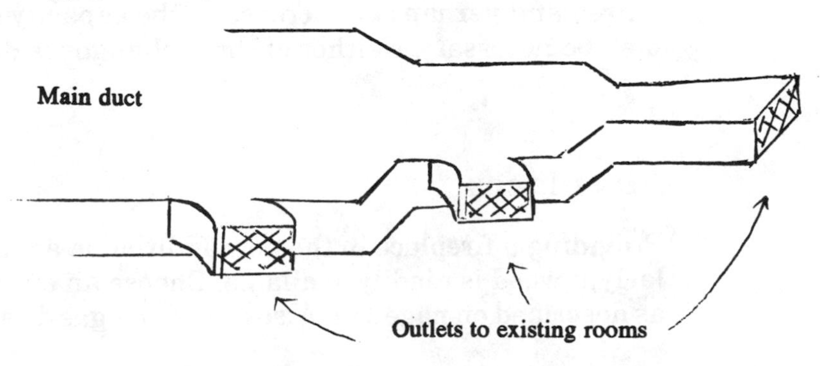

A more efficient solution is to install a through-the-wall heat pump for the new room. The modern version of these systems is very efficient and can be fitted neatly anywhere in the exterior wall.

If air conditioning is not required, an alternate solution for heating is to install electric-resistance heating coils in the ceiling or along the baseboard. Since heat tends to rise, the baseboard coils will most likely produce the more comfortable system.

With this form of additional heat, a thermostat is installed in each new room to control the temperature for that room. If the new room is to be used only occasionally, the temperature for that room can be lowered when not in use. The wall that separates the added room from the original house should be insulated to prevent unnecessary leakage of heat from the original house to the new room when the temperature in the new room is lowered. Also, adding weatherstripping to the door between the new and the old will improve the overall system.

Steam and Hot-Water Heat

If the current heating system is the steam or hot water type, it might be capable of extension into the new addition. Have a professional heating contractor examine the furnace to determine whether it can take on the additional capacity. In addition have the pipes delivering the hot water or steam checked to see if they can be extended. If the original system is very old, the pipes may be clogged and unable to support the original house, let alone any additional system. Under these circumstances, consider replacing the whole system with a new one.

Electric Resistance Heat

If the original heating system is non-ducted electric resistance heat, it probably can be extended easily. Additional circuits may be required and perhaps an increase in the capacity of the service panel may be necessary. Neither of these changes is difficult to make.

Fireplaces

Providing a fireplace in the new addition is another choice, particularly if wood is readily available. Choose an energy-efficient system as described on page 114. Also consider a gas-fired fireplace or stove.

If gas is not available at your lot, you can provide it by using a gas storage tank.

Depending on the weather in your area, a fireplace may do the job by itself or work in conjunction with another system such as electric-resistance heat.

Plumbing

Review Chapter 15, "Plumbing." Usually, the current water supply to your house can take on additional requirements, but the system within your house may need to be checked, particularly if you are adding a bathroom or kitchen. Have a plumbing contractor inspect the water supply pipes to make sure they can take the additional load. If they are too small or internally corroded they should be replaced.

The sewer capacity should also be checked. Usually there is no problem if your house is connected to a public system. If you are hooked up to a septic tank system, however, additional sewage disposal may create a problem if your present tank is too small or the drain field inadequate to take on an additional load. Larger or additional septic tanks can be installed, however, and drain fields can be expanded if your lot has the additional room with proper soil for these fields.

If you are adding plumbing to a basement addition, you may need to have a grinder pump installed to draw the sewage from the basement level to the sewer line if this line is above the floor level of the basement.

Electrical

Usually there are no major problems installing electrical circuits and equipment. Installing additional circuits and increasing the capacity of the service panel are routine electrical tasks.

Financing

Review Chapter 2, "Contracts, Financing and Insurance."

In addition to the usual first mortgage on a home, home improvement costs can be financed by another method—the "equity loan" or second mortgage. If you have lived in your home for some time, you have increased the equity in your ownership. For example, assume that you bought your home a few years ago for $100,000, made a down payment of $10,000 and financed the payment of the remaining $90,000 by a first mortgage. Several years later the market value of your house has probably increased to, for instance, $115,000.

Your present equity is now the $10,000 down payment plus the reduction (assume $2,000) of the principal payments you have made during the last few years plus the $15,000 increase in market value for a total of $27,000. Many banks will lend you up to 80% to 90% with your equity as collateral. This equity loan or second mortgage is paid off in monthly installments along with the monthly payment of your first mortgage.

Sample Specifications

This appendix contains a sample of the specifications for a typical rancher built on a crawl space with an attached garage. The quality of the materials is above average, energy conservation is emphasized.

The owner must provide sufficient detail in the specifications to be sure of getting the desired results and to provide an accurate base for real competitive bidding. If specifications are vague or contain little information, the owner will not be able to compare the prices of the different bidders with any assurance of accuracy.

Although products of certain manufacturers have been used in developing these specifications, this use should not be construed as a recommendation for one certain product instead of any other. By specifying model numbers and manufacturer, the homeowner informs the bidders exactly what he or she wants in the house. The term *or equal* indicates that bidders may substitute products of other manufacturers of the same or better quality and performance.

In certain cases the owner may need assistance in preparing technical details. Most suppliers and contractors are willing to assist in this task without charge.

The paragraphs in italics are not part of the specifications. They are notes explaining to the reader the importance of the preceding specification.

Specifications

Home of Mr. and Mrs. J. L. Brentwood to be built on Lot #29, Jasmine Hills Subdivision, Center City, North Carolina. The owner has the

final determination of acceptable material, equipment and quality of workmanship.

1. **EXCAVATION: Bearing soil for footings to be undisturbed soil.**

   Designates the soil conditions for the footings. Important that the soil not be loose fill.

2. **FOUNDATION**

   a. **Footings: Per plan. Concrete mix 3,000 psi. Reinforcing bar #4. Top of footings to be not less than 8″ below grade.**

      The blueprint will show size of the footings. Necessary to designate type of concrete for strength and to indicate the size of the rebar. The blueprint should show the amount of the rebar. If it does not, that information should be indicated here. In cold areas, specify that the top of the footings should be not less than 8″ below the frost line.

   b. **Foundation wall: Exterior walls to be 8″ concrete block per plan with brick veneer. Interior walls to be 8″ concrete block per plan. Girders and sill plates to be #2 grade SYP salt pressure treated to .25 CCA. Anchor bolts ½″×12″—6′ o.c. Brick to be Triangle Brick Co. 3700 Castletone laid with V–joint and standard mortar.**

      These are important details of the construction of the foundation. The wall description reinforces data that should be part of the blueprint and is repeated for emphasis. The type of lumber and the salt treatment ensures the use of strong, long-lasting lumber for the part of the framing that may be exposed to more than the normal amount of moisture. .25 CCA designates a degree of impregnation normally used for salt-treated lumber installed above the ground. The sizes of the girders and sill plates are shown on the blueprint.

   c. **Termite protection: Soil poisoning by treatment along exterior and interior walls and piers with 1% chlorodane solution.**

      The chemical treatment is the most effective against termites and other wood-boring insects.

3. **FIREPLACE**

   a. **Wood-burning Heatilator Model FP36 with FK15 fan kit, AK20 outside air kit and CD10 heat circulator ducts and outlet boxes, or equal.**

This equipment will provide an energy efficient fireplace as described in Chapter 10. By selecting a specific model with attachments, you express exactly what you want and provide a specific base for bidders.

b. Chimney, per plan, zero clearance Heatilator. Exterior section framed and covered with redwood lap siding to match the exterior of the house. Top of framing covered with galvanized sheet metal cap to prevent moisture entrance into the framing.

Tells the carpentry crew that they must furnish the labor and the supplier the material for finishing off the chimney exterior. Also tells the Heatilator supplier to include the metal chimney sections in the bid.

c. Facing: Colonial wood mantel with ceramic tile per plan.

Detail of fireplace finish to show the carpentry crew, the supplier and the tile contractor the work and material required.

4. EXTERIOR WALLS

a. Studs and plates, SPF KD #2 grade, 2×6—24″ o.c.

Indicates the grade and type of lumber to be used. SPF is a more stable wood than SYP and will provide a straighter wall.

b. Let-in corner bracing, 1×4 SYP at all exterior corners and all intersections of exterior and interior walls.

Sheathing to be used is polyurethane or polystyrene, which has little strength. Diagonal corner bracing is the only practical method to use with this sheathing.

c. Sheathing to be polyurethane or polystyrene 1″ thick 4′×8′ sheets to be applied by hand nailing only. Sheathing to be covered with plastic house wrap.

This material will give much better insulation than either plywood or asphalt impregnated sheathing. The large sheets reduce the seams and make the anti-air-infiltration job easier. Hand nailing will minimize damage to the sheathing. The plastic house wrap will reduce air infiltration through the walls and eliminate the need for some of the other sealing measures. See paragraph 25.

d. **Siding to be colonial beaded 1×6 redwood lap with ⅝″ butt and 4″ exposure. Corner boards, same material, to be 4″×5⁄4″.**

Designates the type, form and size of the siding. The ⅝″ butt is important since it will minimize splitting problems, which would be more prevalent with a ½″ butt. The size of the corner boards was selected for appearance and to provide enough thickness so that the caulking between the corner board and the siding will have a good base.

e. **Gable wall construction to be the same with wood gable vents per plan.**

Because the attic space behind the gable wall is not heated, some saving might have been realized by substituting impregnated sheathing in lieu of the poly sheathing. Since impregnated sheathing is usually available only in ½″ thickness, the poly was used to retain the 1″ depth and to eliminate problems in siding installation because of the different thicknesses of the sheathing. A double layer of the ½″ impregnated sheathing could have been used, but the increase in material and labor costs would have negated any savings.

f. **Exterior painting: All exterior wood to be back primed with one coat of SWP Promar exterior wood undercoat or equal. All exposed surfaces to be painted with one coat of Promar undercoat and two coats of SWP A–100 latex gloss or equal. Rate of coverage to be not more than 400 square feet per gallon for both the undercoat and the A–100. All paint to be delivered to site in its original containers. Colors to be selected.**

Both the quality of the paint and the number of coats is designated to provide a basis for painting bids. The original containers statement minimizes the possible switch to cheaper material. The rate of coverage ensures that the paint is not overly thinned.

5. **FLOOR FRAMING: Joists to be #2 grade KD SYP with blocking not greater than 8′ o.c.**

Designates the grade and type of lumber to be used. SYP was selected because of its strength. Blocking is necessary to prevent undue warping of the lumber. Blocking is more rigid than bridging. It also pulls the joists into alignment before the subfloor is installed. KD indicates that the lumber must be kiln dried rather than air dried. The latter may have an unusually high moisture content with more shrinkage after the house has been built.

6. SUBFLOORING

   a. ½″ × 4′ × 8′ flake board glued and nailed or stapled to joists. Glue to be Weldwood Sub-floor and Construction Adhesive or equal.

      Specifies type and size of material. Should be not less than ½″ and need not be thicker than ⅝″. Gluing increases the strength of the structure and helps to eliminate squeaks in the floor. Stapling has been included to permit the use of powered nailing and stapling systems. Flake board resists moisture much better than exterior plywood and will generally eliminate delamination caused by rain and snow before the house is weathertight.

   b. Attic space: 10 sheets ½″ × 4′ × 8′ sheets of CDX plywood or flake board to be laid near pull-down stairs. Installed only after the ceiling insulation has been put in. Pull-down stair to be Besseler Model 100 size #2 or equal.

      Provides some storage space in the attic. Depending upon the roof construction it is very difficult or even impossible to move full sheets of plywood up the opening of a pull-down stair. This plywood should be placed in the attic space during the framing, but not nailed down until after the insulation has been installed.

7. INTERIOR PARTITION FRAMING: #2 grade KD SPF studs and plates. Studs 16″ o.c. Interior walls of both baths to be staggered 2×4 studs with 2×6 plates and double plaster lathing on one side. 3½″ batt insulation without vapor barrier woven between studs horizontally.

   Staggering the studs and installing insulation provides sound insulation for the bathroom.

8. CEILING AND ROOF FRAMING: Joists and rafters #2 grade KD SYP.

9. ROOFING: Sheathing ½″ × 4′ × 8′ CDX plywood. Shingles 215# fiberglass self-sealing Certainteed Glasstex or equal. Underlayment to be single layer 15# asphalt felt building paper.

   The ½″ plywood is preferred since thinner material may sag between roof rafters, giving the roof a wavy appearance. Heavier plywood is not necessary for this type of shingle. Designating a specific shingle determines the quality and type. Flake board could be used for roof sheathing but it is more slippery than plywood and may increase roof labor costs. Since roof sheathing is usually covered with waterproof building paper right after being installed the plywood is normally dry and not subject to delamination from rain and snow.

10. GUTTERS AND DOWNSPOUTS: Per plan. Material to be Raingo All-Vinyl Rainwater Handling System or equal. Concrete splash blocks at the bottom of each downspout.

    Blueprints usually show the gutter and downspout detail. If they do not, the lengths and number of downpspouts and length of gutter should be determined. If needed, get some help from a supplier. A particular brand is specified to establish quality, in this case a tough vinyl type. The splash blocks (usually formed concrete) reduce the erosion of the newly graded yard.

11. INSULATION

    a. Ceiling R–30 with vapor barrier 6-mil polyurethane under plaster lathing.

    b. Walls R–19 with vapor barrier 6-mil polyurethane under plaster lathing.

    c. Floor R–13 with vapor barrier of kraft paper.

    d. Interior walls of each bath 3½″ batts without vapor barrier for sound insulation.

       The polyurethane vapor barrier is more effective than the integrated barrier in the batt. The sheathing and plaster will raise the total R–value of the walls to about R–19.

12. LATH AND PLASTER: All walls and ceilings to be two-coat plaster veneer with gypsum lath base, including the interior of all closets and the garage. Lathing applied to studs and joists with adhesive and screws except for exterior walls, where screws only are to be used.

    Method of installing lathing is detailed to eliminate the use of lath nails and reduce the use of screws. Nails have the tendency to pop out from the studs. Adhesive cannot be used on exterior walls because of the polyurethane vapor barrier.

13. DECORATING

a. All walls and ceilings except kitchen and baths to be painted with one coat oil based primer and one coat of SWP Classic 99 latex or equal.

Quality and number of coats specified.

b. Walls and ceilings of kitchen and baths and all-wood trim (including doors and windows), one coat of oil-base primer and two coats of SWP Classic 99 latex enamel semigloss or equal.

Use enamel paint in rooms with a higher than normal moisture content and on all wood trim.

c. Rate of coverage of all painting to be not more than 400 square feet for each gallon.

14. INTERIOR DOORS AND TRIM

a. Doors to be six-panel colonial, wood, paint grade, Morgan F–66 or equal.

b. Door and window trim to be colonial 3¼″. Window trim to be installed "picture frame" without stool and apron.

c. Base trim to be colonial bead 4¼″ with shoe molding where vinyl, earthstone and hard wood flooring are laid.

Type and quality of doors has been established. Size and style of trim and method of trimming windows is specified.

15. WINDOWS: Size per plan. Pella clad casement windows, dual glazed, dark bronze with removable colonial muntins and complete operating hardware. Slim Shades installed in all windows in rear of house in lieu of muntins.

The style, color and type of window is established, as is the quality. "Slim Shades" in the Pella line are venetian blinds installed between the two glass panes. Muntins are removable strips that give the appearance the window is made of small glass panes—the traditional look. Both slim shades and muntins fit between the panes of glass, therefore, only one or the other can be used in the same window. The interior pane of the Pella window is removable, giving access to the muntins and the Slim Shades.

16. ENTRANCE DOORS

a. Front entrance door to be Benchmark insulated steel door and frame, Williamsburg model, six panel or equal. Size 3′0″×6′8″. Hardware to be Weiser E360DL Troy, finish 12.

b. Kitchen to garage door to be Benchmark insulated-steel door and frame, Willliamsburg model, six panel or equal. Size 2′8″×6′8″. Hardware to be Weiser E360DL Troy, finish 12.

c. Sliding door in family room to be Pella clad, dark bronze XO 34-8 dual glazed with sliding screen door or equal.

The above specifications establish the size, quality, finishes and style of all exterior doors and door hardware.

17. CABINETS

a. Kitchen cabinets to be Wheaton Oak by American Woodmark or equal. Countertops to be plastic laminate, postformed.

b. Bath vanities to be Nutone Savannah or Equal.

The range of type and quality, both which reflect price, seems to be endless in the cabinet line. Select a specific line you like and use it as the requirement in your specifications. The blueprints will show size and arrangement of cabinets. Designate the type of countertop as they differ in price. See paragraph 21 g for countertops for vanities.

18. FINISH HARDWARE (Medicine cabinets, towel racks, door locks, etc.): Allowance for material—$300.

By using an allowance the owner provides the cost of material and the contractors bid for labor only. This method permits the owner to postpone the selection of the items until construction is underway. Do not delay too long, however, since the framing crew must know the rough opening for the medicine cabinets and the location of other hardware so that any required framing back-up can be installed. The labor for installing the finish hardware is normally included in the labor for interior trim.

19. FLOORS

a. Entry hall to be earthstone set in cement over dropped subfloor per plan.

b. Kitchen and laundry to be Congoleum Prestige Cushioned vinyl or equal installed over ⅝″ plywood underlayment.

c. Bath floors to be ceramic tile American Olean C1 crystaline face, or equal, installed over cement on plywood subfloor. Base to be matching ceramic material.

Be specific in your selection of tile. Prices of tile vary. Detail the method of installation. The one used here is a good, trouble-free one.

d. Wood flooring in living and dining rooms. To be select white oak random width 3″, 4″, 6″ with V–groove. Finish to include sanding, one coat of stain and two coats of polyurethane, matte.

Typical example of the detail needed in wood flooring.

e. All other flooring to be carpet installed over ⅝″ particle board. Allowance of $14 per square yard installed for the carpet.

20. WAINSCOTING: Tubs in each bath to have ceramic tile laid over tub on metal lathing and cement to height of 6′ from floor. Tile to be American Olean A–1 matte glace or equal.

21. PLUMBING

a. Water supply: hookup to public system.

b. Sewage disposal: hookup to public system.

c. House drain inside and out to be PVC. All vent pipe will exit through the roof on the rear side of the house.

The house will look better from the front if vents are on the rear side of the roof where they cannot be seen from the street.

d. Water supply pipe, both hot and cold, to be polybutylene. All hot water pipe run through exterior walls and in unheated spaces will be wrapped with 1″ pipe insulation.

e. Sill cocks: Two per plan.

f. Water heaters: Two Ruud #PE40-2 40-gallon Energy Miser or equal. Installed per plan, one in vicinity of bath area and one in vicinity kitchen/laundry area. Both hot water heaters to be wrapped with additional insulation of not less than 2″ of fiberglass.

Denotes the quality and capacity of the hot water heaters. Illustrates the use of two heaters for more efficient distribution of hot water. Extra insulation is specified because both hot water heaters will be installed in unheated attic space.

g. **Fixtures:**

Bath #1: American Standard water closet 2109395 in blue. American Standard 14″ cast iron tub 2265379 in blue with Moen 2434 tub/shower combination controls. Mount controls halfway between shower head and tub supply outlet. Corian double integrated bowl lavatory with two Moen 4220 faucets. Shower rod. Or equal.

Bath #2: American Standard water closet 209395 in yellow. American Standard 14″ cast iron tub 2265379 in yellow with Moen 2435 tub/shower combination controls mounted halfway between shower head and tub outlet. Corian double integrated bowl lavatory with two Moen 4220 Faucets. Shower rod. Or equal.

Kitchen: Elkay CR 3322 stainless steel sink with Moen 73-10 faucet with spray. Or equal.

Laundry: Fiat F–1 laundry tub with Sterling 31–200 faucet. Or equal.

Since there are many grades and styles of bath fixtures, you should indicate manufacturer, model and color. The mounting of the tub / shower controls is spelled out since this is the most convenient position for the user. Otherwise the plumber may mount the control too low for convenient operation of the shower.

h. Hookup for garbage disposal and dishwasher.

i. Install Water Tite washer box in laundry for clothes washer.

"Water Tite" is an in-the-wall box that provides an efficient, nice-looking recess for the hot and cold water valves and the drain.

22. HEATING, VENTILATING AND AIR CONDITIONING (HVAC)

a. Carrier 38Q046—4-ton heat pump and matching fan coil with 15KW auxiliary strips. Thermostat with separate outdoor thermostat to stage electric heat strips. Minimum HSPF 6.5, SEER 10.

Specify the manufacturer, model number, capacity and additional features to ensure that you get the proper system. Since these specifications are somewhat technical, contact a contractor and get some help. The separate outdoor thermostat is important since it will bring on the auxiliary heat in time to avoid a temporary drop in the inside temperature.

b. **Humidifier integrated into system to be Aprilaire Model #440 bypass type.**

Forced-warm-air heating systems will dry out the air inside the house. The humidifier will put moisture back into the air and provide greater comfort for occupants at lower temperatures and reduce damage to furniture caused by excessive dryness.

c. **Electrostatic air cleaner to be Carrier unit 31MP220 integrated into the forced-warm-air heating system.**

Provides cleaner air inside the house. This is particularly important in a well-sealed house.

d. **Ducts to be galvanized sheet metal with complete sound insulation lining throughout. In addition to the sound insulation, one layer of 2″ duct insulation applied to the exterior of all ducts passing through unheated areas.**

The lined ducts will provide a quiet system. The lining also offers additional thermal insulation. As indicated, add insulation to unheated areas.

e. **All duct openings, both supply and return, will be covered temporarily after rough-in to prevent debris, dust and other items getting into the duct system. Covers to be removed when grilles are installed during final.**

23. ELECTRICAL WIRING

a. **Underground service with 200-ampere circuit breaker panel with 24 circuits.**

Underground service eliminates the unsightly wires coming into the house from the power company's transformer. The size of the panel and the number of circuits control the power your house can handle.

b. **Wiring to be nonmetallic copper cable except that aluminum–stranded cable may be used for #8 and larger.**

Standard for house wiring. The "non-metallic" refers to the cable cover and not the copper.

c. **Special outlets for range, water heaters, oven, clothes dryer and heat pump.**

Highlights the special 220-volt circuits needed for this equipment.

d. **Door chimes with button at front door.**

24. **LIGHTING FIXTURES: Allowance of $600 to include electric wall heaters for the bathrooms and ceiling fans in the family room and master bedroom.**

    The allowance method permits the owner to select fixtures at a later date.

25. **SEALING**

    a. **Door frames in exterior walls and all window frames to be sealed to rough openings.**

    b. **All electrical, plumbing, telephone and heating penetrations of exterior walls, floor and ceiling of heated space to be sealed.**

    c. **All sheathing construction damage including nail holes, breaks and utility openings to be sealed.**

    d. **All cracks in framing members where daylight is visible from the inside to be sealed.**

26. **SPECIAL EQUIPMENT: Allowance of $2,500 for appliances.**

27. **SIDEWALKS: Per plan 4″ thick exposed aggregate 3,000 psi with wire-mesh reinforcing. 3′ wide mark.**

28. **DRIVEWAY: Per plan. 6″ compacted clay base with top coat of 2″ of gravel. Culvert to be 24″ flared-end concrete pipe.**

29. **GARAGE DOOR: Wayne overhead 16′×7′ model 54 with Genie Model #880 by Alliance Mfg. Company radio-controlled operation with two transmitters. Or equal.**

    Specify the type of garage door. Since this is a double garage, two transmitters have been specified for the two cars that will use the garage.

30. **LANDSCAPING: Final grade with slope away from house foundation. Grade, fertilize and grass seed all of the area of the lot that has been cleared for construction. Lay cover of straw over the seeded lawn.**

3. Act as a guide for installing the batter boards after the lot has been cleared

To start this location process, make a plat (to-scale drawing) of your lot to include the setback from the lot boundaries required by local codes or covenants. In the example shown in Figure C.1 the setback required for the front of the lot is 30′, the sides 15′ and the rear 20′. The house must be built within this area.

In the example shown, the 64′-6″ × 42′-0″ rancher is to be placed 45′ back from the front of the lot and 60′ from the right boundary.

To square up the house location, know the diagonal dimension of the house, the distance from stake 1 to 4 or from 2 to 3. This diagonal is computed by taking the squares of the two sides, adding them together and determining the square root of the result. In this case the figures would be 42 × 42 plus 64.5 × 64.5, which equals 5,924.25, the square root of which is 76.969 or 76′ 11⅝″.

To locate stake 1 measure 45′ in from the front boundary and 60′ in from the right side boundary. Place the stake at this junction. Measure 64′ 6″ to the left and place stake 2 in the ground. Check the location of stake 2 by making another measurement of 45′ from the front boundary. To facilitate the location of the other two stakes, place a small nail in the top of stakes 1 and 2. Take two tapes (use 100′ ones) and hook one of the tapes over the nail on stake 1 and the other over the nail of stake 2. Pull the two tapes together at the vicinity of stake 4's position with a reading of 76′ 11⅝″ for the tape anchored at 1 and a reading of 42′ for the tape anchored at 2. The intersection of these two tapes at those readings establishes the proper location for stake 4. With the tapes still hooked on stakes 1

FIGURE C.1 Plat

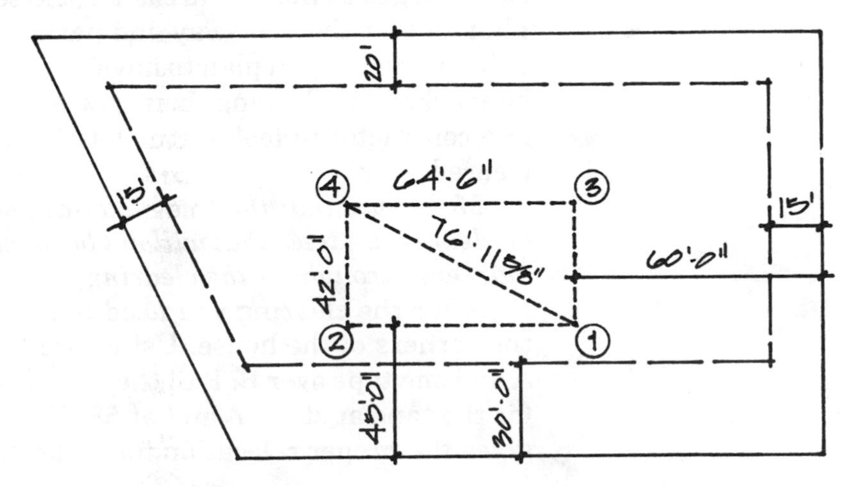

and 2, reverse the readings and pull the two tapes together at these readings to accurately locate the position for stake 3.

As a double check, remeasure the front and back dimensions of the house. If they are within about 2″ or 3″ of what they should be, your temporary location is OK. Check the two diagonals. If they are the same and read close to 76′ 2⅝″ your diagram of the house is square.

If clearing of your lot is not necessary, skip ahead and start the work described in the section describing the batter-board system, which begins on page 259.

Marking the Lot for Clearing

If you must clear your lot of trees and brush, you must take an interim step before the batter-board system can be installed. If you leave the temporary stakes in place, they will undoubtedly be rooted out by the lot clearing machinery; steps must be taken to save the work already done.

Put two stakes A and B in the *ground inside the area that is not to be cleared.* See Figure C.2.

They should be 50′ or more apart with a clear line of sight to the two front stakes of the house, 1 and 2. If it is more convenient, the rear or side stakes can be used instead of the front stakes.

Measure the distance from each stake A and B to each of the stakes 1 and 2 and record these distances. You now have a means of relocating the two front corners of the house later.

Having laid out the house, the next step is to specify what area you want the site work contractor to clear. Using a colored plastic tape (brilliant orange or red is fine), mark the trees all around the house about 10′ away. You will need this extra room for construction. Don't forget to include areas for the septic system if one is in your plan and for the driveway and parking areas. This is the minimum to be cleared. If you plan to have lawns, then additional space should be marked for clearing. It is now a relatively simple matter for the site contractor to look at the lot and understand just what is to be cleared.

Make sure that the trees you have marked to indicate the area to be cleared are trees that will not be taken out. You want to retain your markers throughout the clearing job.

After the clearing you need to relocate the four stakes showing the corners of the house. Using two long tapes, hook one tape over A and one tape over B. Pull the two tapes together at the 95′ 8″ mark for the tape at stake A and at 58′ 7″ for the tape at B. This junction gives the proper relocation for stake 1. Do the same to locate stake

FIGURE C.2 Temporary Layout before Clearing

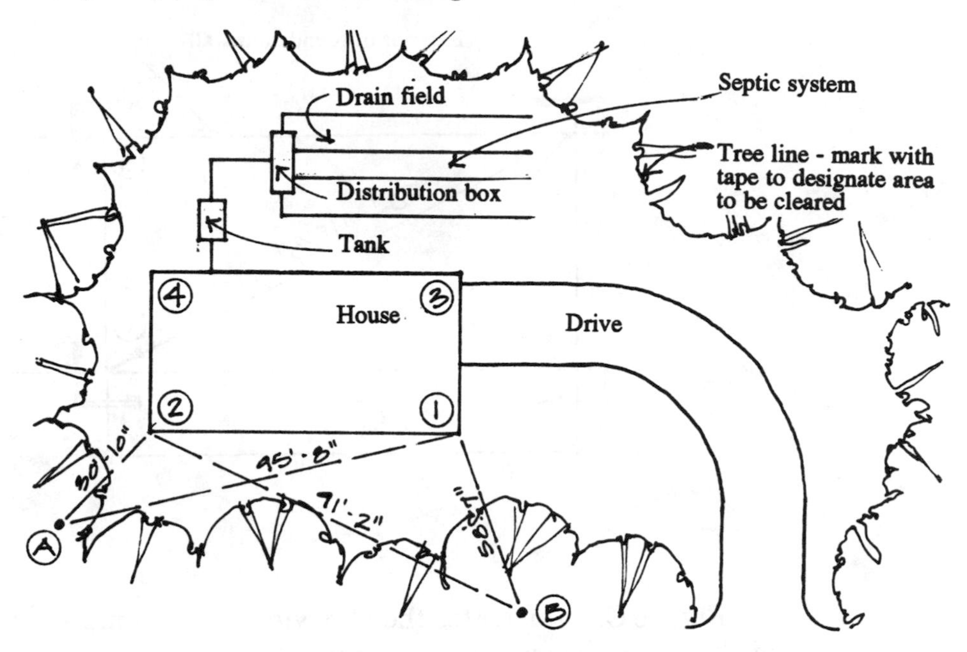

2 by intersecting the two tapes at the 30′ 10″ mark for the tape on stake A and the 91′ 2″ mark for the tape on stake B. You now have the front of your house relocated. With this front line as a base you can locate the remainder of the house using the system previously discussed.

The Batter-Board System

The purpose of the batter-board system is to establish the exact location of the following elements:

1. The outside top edge of the foundation walls
2. The centers of internal foundation piers
3. The height and levelness of the floors of the house

In other words, the batter-board system establishes the base for construction of the house. It must be installed accurately. Carelessness can cause mistakes that could be very expensive to correct. The system picks up from the temporary stake locations discussed previously.

FIGURE C.3 Batter-Board Plan

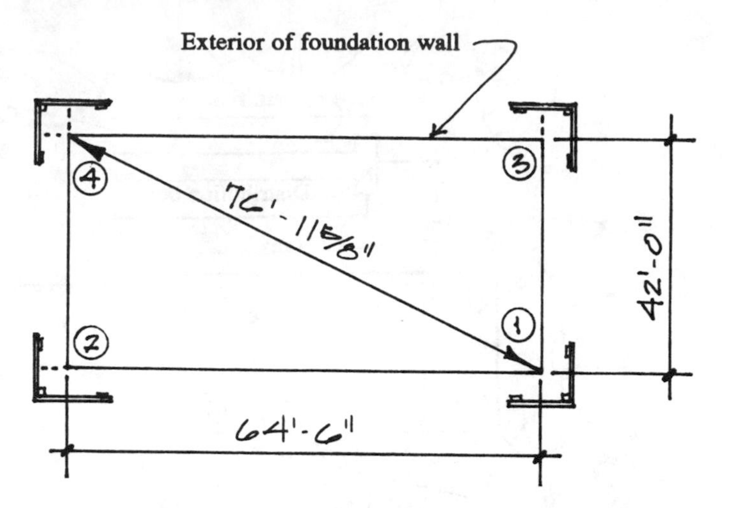

Figure C.3 illustrates the plan view of the completed installation of the batter boards for the sample house.

The composition of the batter-board structure for foundation corners and for junctions of foundation walls is shown in Figure C.4. The application of the junction or single batter board for the L-shaped house is discussed later in this appendix.

The vertical stakes for the system should be about 4′ long and of 2×4 material or roughcut 1×4. These sizes will suffice for a relatively flat lot. If your lot has considerable slope, you will need stakes longer than 4′ and the structure may require bracing to retain the necessary stability. The horizontal boards should be 1×4s, either roughcut or finished lumber, and they should also be about 4′ long. Each corner batter-board system requires three stakes and two horizontal boards. The single-string batter-board system requires only two stakes and one horizontal board.

For the corner batter boards, drive three stakes in the ground so that when the horizontal boards are installed they will be roughly at right angles to each other. This angle is not critical. Keep the base of the stakes at least 2′ from and roughly centered on the small stake you previously put in the ground. If the batter-board stakes are too close to the corner, they may be disturbed by the excavation for the footings and then you have lost the base for determining the exact corner of the house. Drive the 2×4 stakes in the ground using a sledge hammer of six pound weight or more. Keep your eyes riveted to the top of the stake to avoid misses. Drive the stake into the ground far enough to give it a good solid feeling.

FIGURE C.4 Batter-Board Details

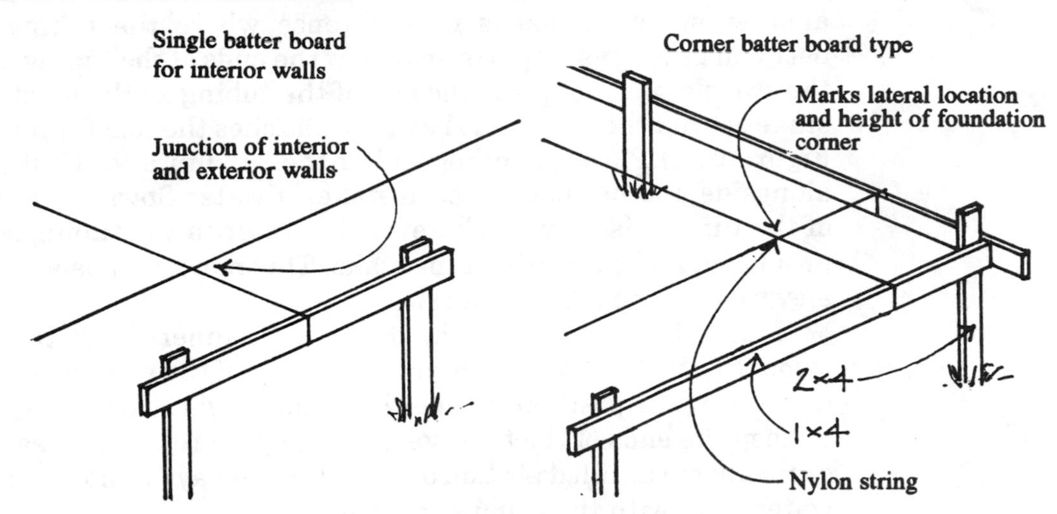

The next step is to establish the horizontal layout. Beginning at the stake that is on the *highest* piece of ground of the corners, mark on this stake the minimum clearance for the crawl space required by the local building code. Add 3″ or 4″ to make sure that your crawl space clearance meets the code.

This mark establishes the height of the foundation walls. Now you must transfer this same level to the batter-board stakes for the other three corners. This will ensure that the first floor of the house is level.

There are two common systems for accomplishing this task. One is the use of a level or transit/level. If you have access to one and already know how to use it, then go ahead and get the job done. If you do not know how to use one of these instruments, then switch to the system described in the following paragraphs. It would take too many pages and illustrations to include instructions on the use of the transit in this book.

The second method of marking the batter-board stakes is simpler and more foolproof. You will need a length of clear or translucent plastic tubing about ⅜″ in diameter and long enough to reach from the stake with the master mark to each of the other stakes. In the example layout, you should have about 90′ to span the diagonal with enough slack to do the job. You will need two clips with which to close each end of the tubing when it is not being used. Fill the tubing with water, leaving about a foot unfilled at each end and clamp the ends to retain the water. Those heavy-duty black spring clips used in most offices do a fine job.

Stretch the tubing from the stake with the master mark to another stake. It makes no difference where the tubing lies in between as long as all parts are below the ends of the tubing. Remove the two clamps and place the end of the tubing at the master mark stake so that the water level exactly matches the master mark. (See Figure C.5.) Place the other end of the tube in a vertical position alongside another batter-board stake. If water flows out of the end of the tube, raise it up. Wherever the water in the tubing seeks its own level, make a mark on the stake. This mark will be at the exact elevation as the master mark.

Mark the other stakes in the same manner. With the clamps removed, the level of the water at each end of the tubing as it is held in the vertical position will be the same height. Therefore, be sure to keep the ends of the tubing high enough in relation to each other so that some unfilled portion of the tubing always remains above the water level with the clamps removed.

The purpose of the marks you have put on the stakes is to determine the location of the top of the batter boards when they are attached to the stakes. Place the mark on the rear of the stake because that is the side on which you will nail the batter boards.

The next step is to attach the batter boards to the stakes on the rear side of the stakes so that the top of the batter board is exactly in line with the level mark. You will need the sledgehammer used to drive the stakes, some eight penny nails or larger and a hammer. It is easiest if three people do the job: one to hold the sledgehammer as a backstop for the nailing, one to hold the batter board and ensure that it remains in the position right at the level mark during the nailing and the third to do the actual nailing. Now install the batter

FIGURE C.5 Water-Tubing Method

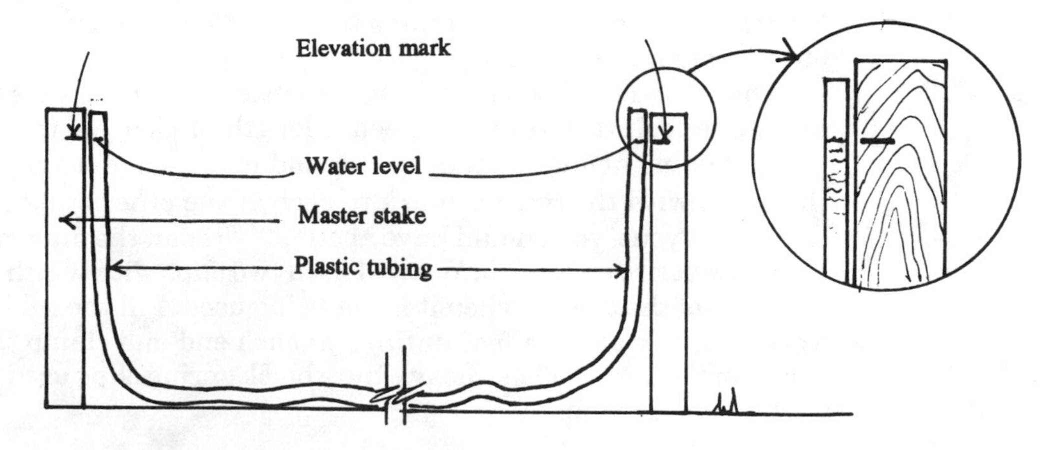

boards. After you have finished, check their level with the water-filled tubing. Strive for a variation not over ⅛″.

To change the elevation of a batter board should it need realignment, use the sledgehammer to drive the stake farther into the ground or use it lightly against the underside of the batter board to bring it out of the ground. If you have done your nailing properly, very little change should be necessary.

So far the only important measurement is the elevation of the tops of the batter boards. The stakes need not be perfectly vertical, nor is it necessary that the batter boards be at true right angles to each other.

The next step is to install the nylon string lines. Use nylon because it is elastic, can be easily stretched taut and is strong—so strong that you should not attempt to break it with your hands.

Run the string across the top of each batter board to its opposite at the other end of the rectangle. A special method for temporarily tying the string is shown in Figure C.6.

In most cases the string will have to be moved a couple of times before it is locked into position. Initially place the string so that it crosses approximately over the small stakes you used to establish the temporary location of the four corners.

The next step is the process of refining the string layout and locking it in place. Select one side of the house as the base line, for example the front shown as line 1–2 in Figure C.3. With the tape, measure along line 1–2. It should be 64′ 6″. *You are not measuring between batter boards; you are measuring the distance between the intersections of the front wall line and the two side wall lines 1–3 and 2–4.* This measurement cannot be made accurately unless you use the batter boards to support the weight of the tape and to let you pull the tape taut. (See Figure C.7.)

You will have to use an artificial zero with this system because the real zero will fall outside the length of string you are measuring.

FIGURE C.6 Temporary Method of Wrapping String

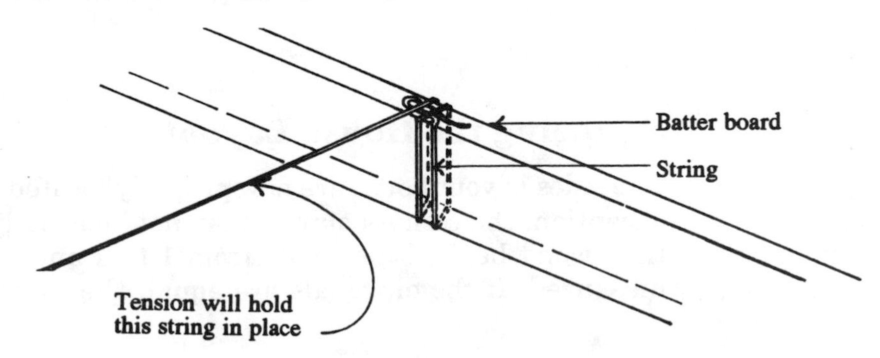

FIGURE C.7 Method of Bracing Tape on Batter Board

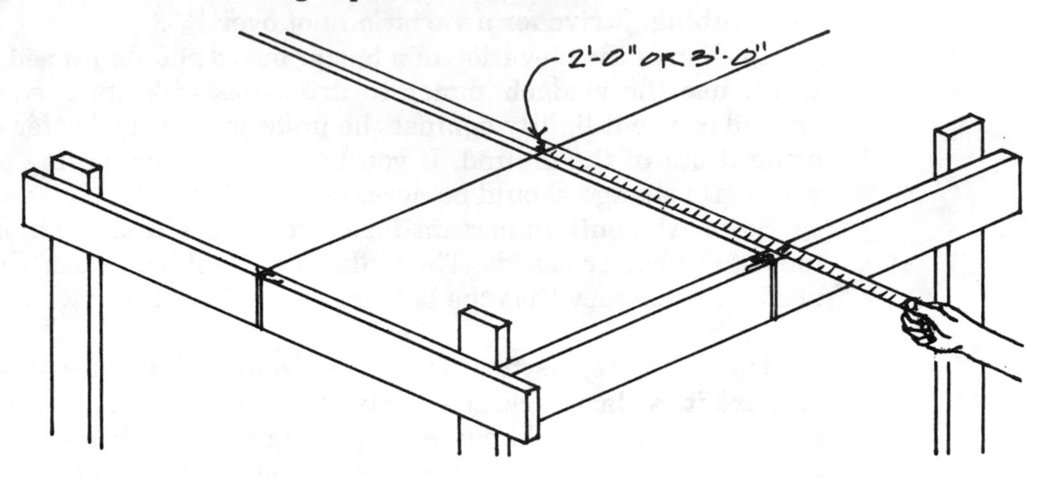

Select an even foot mark such as 2′ or 3′ and subtract this amount from the total measurement. Make all measurements using this system, including the measurement of the diagonals.

In Figure C.3, the distance between the two corners of the front is 64′ 6″. If your measurement is not exactly the same, move the line 2–4 in the proper direction to compensate for any differences. Following the same procedure, measure lines 2–4, 4–3 and 1–3 between the corners of the house, making the necessary movements to correct the readings. Now recheck the whole system, because if you move one end of a line it will probably change the other end of the same line. In any event, remeasure the system until all of the readings, four in this case, come within ⅛″ of the correct reading.

Until it is locked into permanent position, one of the strings may occasionally come loose. The temporary locking method is not foolproof. To prevent having to remeasure the house layout all over again should this happen, use a felt pen to mark the top of the batter board exactly where the string crosses over. Should the string come loose, merely place it back into position across this mark.

Squaring the House Layout

The sides of your house are now properly located with one important exception: the corners may or may not be exactly 90°. The odds are they won't be. The diagonals from 1 to 4 and from 2 to 3 must be measured. If the diagonals are equal, the corners are square, but you may have a trapezoid. If the diagonals are equal and within ⅛″ of 76′ 11 ⅝″, the corners are square.

Let's assume, as is usually the case, that the two diagonals do not match. This shows that the house is definitely not square and the corners not 90° angles. Let's also assume that you have measured diagonal 3–2 as 77′ 5″ and diagonal 1–4 as 75′ 9″. You must shorten diagonal 3–2 and lengthen diagonal 1–4, but you want to preserve as much as possible the lengths of the sides of the house you have already established. To do this, move the 3 end of line 3–1 and the 4 end of line 4–2 to the left (facing the house). See Figure C.8.

Both of these ends must be moved the same amount or you will throw off the dimensions of the sides. In this case the difference between the lengths of the two diagonals is 20″; therefore move each line about 10″. This movement won't always completely correct the differences of the two diagonals but it should be close. Measure the diagonals again. If they are not within the ⅛″ tolerance, then repeat the process until this tolerance is achieved.

Now go back to the beginning and recheck all the dimensions of the sides. Whenever you have to move one line in the checking process, go back and recheck the system until all measurements meet the ⅛″ criterion.

There is one final step before the house layout is complete—locking in the line locations so that they can be removed easily (the footing contractor and the mason will want to do this) and returned to their proper position quickly and accurately.

Two systems are in common use. One method is to drive a nail into the top of the batter board at the proper location and tie the string to this nail. The other is to use a horizontal saw cut into the top of the batter board, as shown in Figure C.9. This is the best method.

After all dimensions have been finalized, mark with the felt pen exactly where the line crosses the top of the batter board. Remove

FIGURE C.8 Moving Diagonals

FIGURE C.9 Permanent Method of Wrapping String

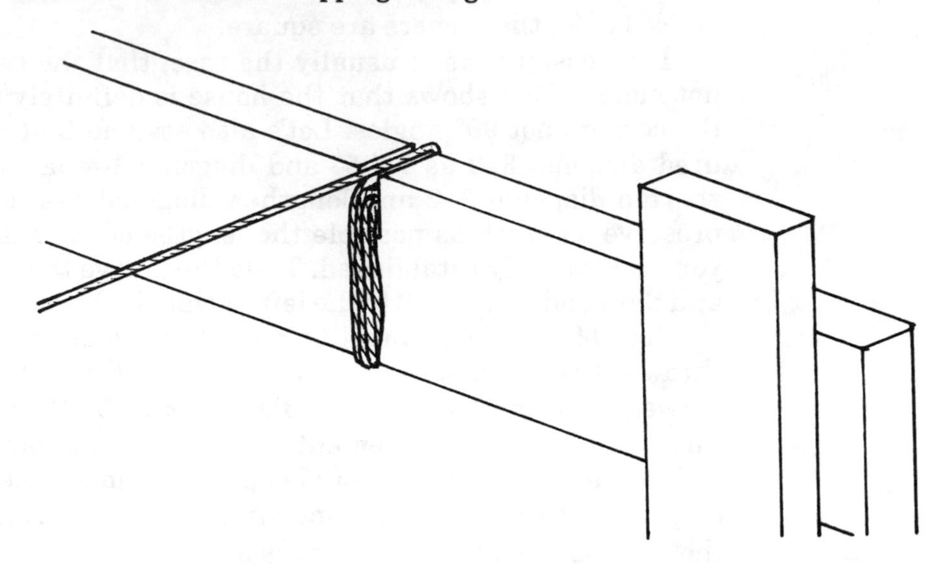

the string and make a saw cut on this mark about the depth of three layers of string. Reinstall the string and pull it very taut. Bring the line across the batter board but not in the slot. Run it to either side of the slot. Continue wrapping the string around the batter board with successive loops going into the slot (at least three). Finally, move the first loop into the slot so that it rests on top of the others. The tension of this loop will hold the line firmly in place and yet make it very easy to remove and replace.

As an additional precaution and as an assist to the mason who will install the concrete blocks, place a large X on the vertical surface

FIGURE C.10 Mason's Mark

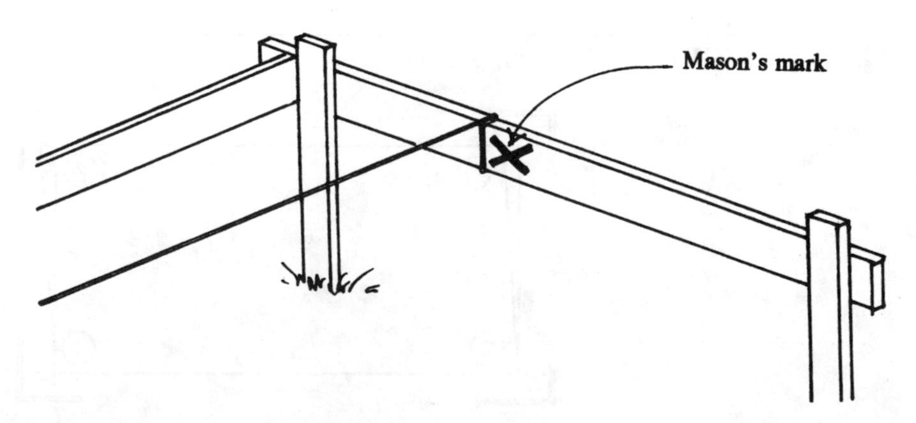

of the batter board next to the string to show on which side of the string the block is to be installed. See Figure C.10. In a house as simple as the one laid out so far, this measure is probably not necessary. In more complex houses, however, particularly those with interior foundation walls, it can save a lot of grief by ensuring that the foundation wall is laid in the proper position.

Layout of the L-Shaped House

The layout of the L-shaped house illustrates the use of the false corner.

In Figure C.11, note that the corners 1, 2 and 3 are actual corners of the foundation, but 4 is not. It is a false corner used to simplify the squaring of the house. The procedure begins with the layout of the four corners 1, 2, 3 and 4 in the same manner described for the rectangular house.

With that step complete, the corners 5, 6 and 7 are added. For corner 5, simply measure the distance 3–5 along the string line and mark the location of 5 with a felt pen on the string. Do the same for 7. Install the single-string batter-board system for both 5 and 7. Install a double or corner batter board for 6. Temporarily put up the

FIGURE C.11 Layout of the L-Shaped House

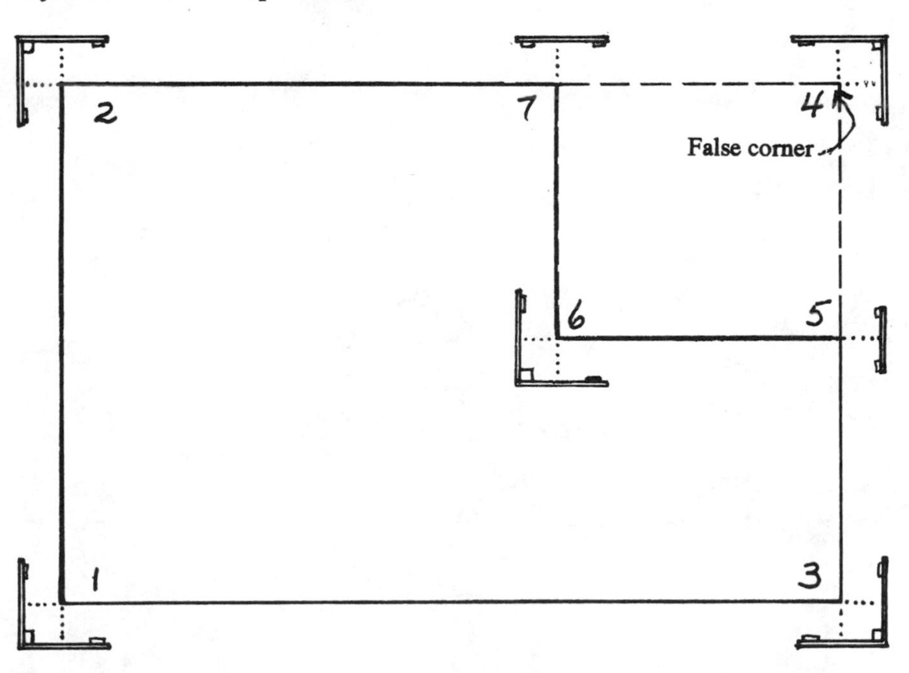

string lines 5–6 and 6–7. Make whatever adjustments are necessary so that line 6–7 equals 5–4 (dimensions taken from the blueprint) and line 5–6 equals 4–7. *Under no circumstances disturb your original layout of 1–2–3–4.*

Since the base from which the new corner was established has already been squared, it is not necessary to square the 5–4–7–6–5 rectangle.

Using the principles illustrated in the rectangular house and the L-shaped house, you should have no problem in laying out any other shape consisting of rectangles such as the U, the H or the T.

House Layout Material

You will need the following:

- Bundle of small stakes 1″×2″×12″
- Batter-board material (1″×4″), stakes (2″×4″), nails and sledgehammer
- Plastic tubing
- Two steel tapes, each 100′
- Pocket steel tape, 10′ or 12′

F

The Superinsulated House

For those who are building in the very cold climates of the northern United States, including Alaska, and Canada, where winter temperatures drop to 20° to 30° F below zero, often with strong winds, special construction techniques are necessary to attain an energy-efficient house.

In these very cold climates, the proper R–values of insulation can be as high as 60 for ceilings (attic), 42 for exterior walls and 36 for the floors. Reduction in heating costs up to 90% may be realized through the following:

1. The complete sealing of the exterior shell against air infiltration by the careful application of plastic house wrap
2. Framing alterations of the exterior walls to provide the depth needed to install up to 12″ of insulation
3. Some roof modifications to permit application of insulation up to R-60
4. Design changes so that most of the windows face the south and all windows are equipped with interior shutters

With the large increases in the cost of energy in recent years, most people now recognize the need for increased insulation to reduce heating and air-conditioning costs. In the superinsulated house, the reduction or even elimination of heat loss by means of the infiltration of air through the exterior shell is certainly as important. The average newly built house has an air leakage area equal to a single hole about 2′ square. Many older homes have as much as four times this amount of leakage. This appendix illustrates these structural techniques necessary to achieve superinsulation.

Single-Stud Wall Framing

The steps for the proper application of an air and moisture sealing vapor barrier are illustrated in Figure F.1.

The steps are carried out as follows:

A. Staple 6-mil polyethylene to the interior side of the studs as shown. Note that all joints between adjacent sheets of poly are made over a solid part of the wall or ceiling such as a stud or joist.

B. Where seams in the poly are to be made, apply a bead of sealant on the top of the poly where it has been stapled. The sealant used must be nonhardening material available for use in cartridges. The most practical choice is an acoustical sealant. Do not use sealants or caulks that tend to harden.

FIGURE F.1 Technique for Joining Vapor Barriers on Stud Walls

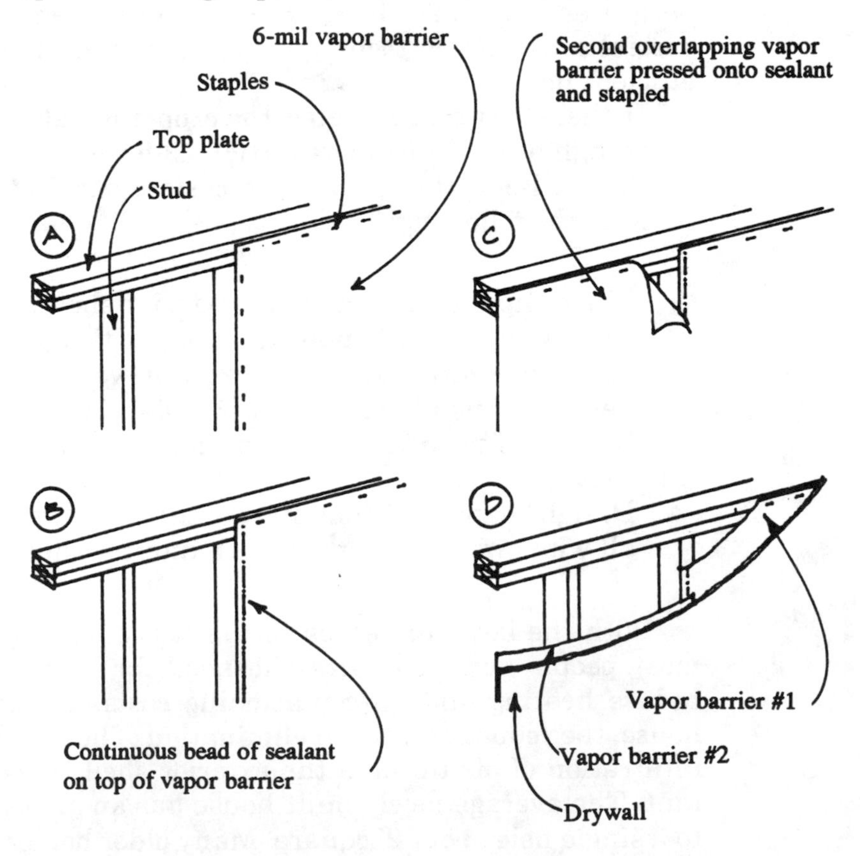

They will eventually crack and expose the joint to air leakage.

C. Staple the adjacent sheet of poly to the studs and plates and, at the overlap seam, press it into the sealant as shown to form an air-tight bond. Make sure that the sealant bead is continuous; there should be no gaps.

D. Cover the poly vapor barrier with rigid material such as plywood, sheetrock or wall paneling. The solid backing of the seams of the poly and the rigid covering are necessary to withstand the forces of the wind, which can pull the sheets apart.

Figure F.2 illustrates a section of the house showing how the vapor barrier is made continuous between the floors and between the wall and ceiling. Air leaks will occur if these areas not sealed as shown.

Figure F.3 illustrates a method of sealing the single exterior wall around electrical outlet boxes. The major point is that a vapor barrier must be installed *behind* the electrical outlet and the insulation. Proceed as follows:

1. First install the wood blocking to isolate the area. This blocking can be 2×4 or 2×2.
2. Next, install the vapor barrier in the blocked off area. This can be accomplished by the use of a commercially available plastic pan that looks much like a roasting pan about 2″ deep, or, more simply, make a pan out of the poly on the job. Be sure that there are no unsealed seams in the job-made pan. Staple the pan to the studs and wood blocking as shown.
3. Drill a hole through the blocking and pan and pass the electrical cable through the hole to the vicinity of the outlet box.
4. Wire the box and seal the cable opening as shown.
5. Fill the pan with insulation. This puts the insulation on the inside of the vapor barrier of the pan.
6. Apply a continuous full bead of sealant around the pan over the staples.
7. As the overall vapor barrier for the full wall is installed, press this vapor barrier well into the sealant.
8. Cut a hole in the overall wall vapor barrier to provide access to the outlet box.

Figure F.4 shows the technique for pipes and vents penetrating the vapor barrier. To avoid problems with freezing pipes and vapor penetration, plumbing pipes should normally be run through interior walls and through floor joists inside the heated area of the

FIGURE F.2 Wall Section: Vapor-Barrier Detail

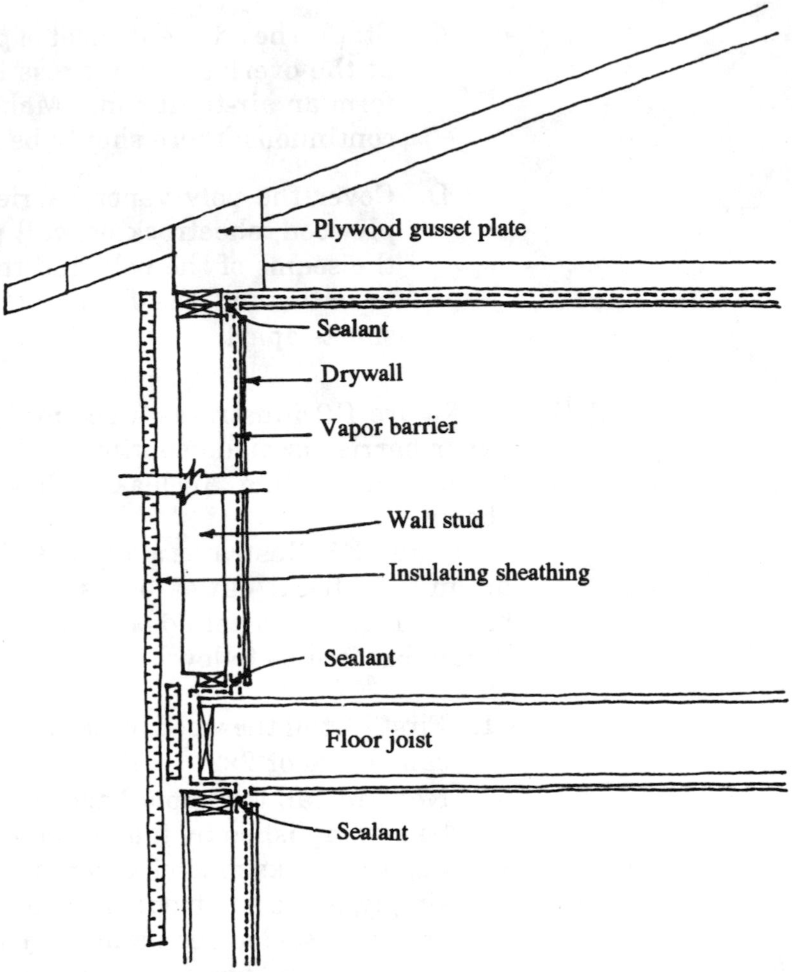

house. Polybutylene pipe (PB) *will not burst when the water inside freezes.* It expands with the frozen water and resumes its normal shape when the water melts. This material should be considered for plumbing water supply (both hot and cold) pipe by anyone building in cold climates.

Double-Stud Walls

In many cases, single-stud walls, even those built with 2×6 or 2×8 studs, do not provide enough space for the required thickness of

FIGURE F.3 Vapor-Barrier Installation at Electric Box

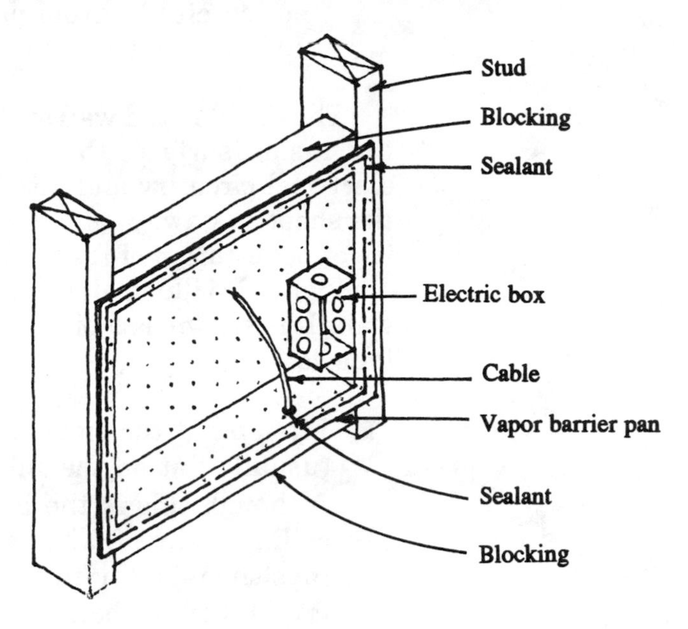

insulation. In addition the single-stud wall has the following disadvantages:

- The vapor barrier may be punctured during construction, thus seriously reducing its effectiveness.
- Unless special measures are taken, the electrical outlets must penetrate the vapor barrier unless they are installed on the surface of the interior wall finish.

FIGURE F.4 Technique for Sealing Penetrations of the Vapor Barrier

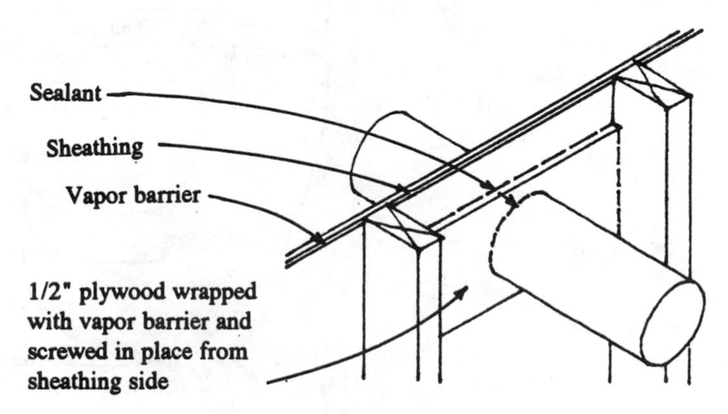

- Installers may ruin the vapor barrier sealing job if they are not careful to avoid penetrating it when applying sheetrock and paneling.

The double-stud wall system places the vapor barrier away from the inside surface, thus providing space for the installation of electrical circuitry and interior insulation. The system also moves the sheetrock away from the vapor barrier and reduces the possibility of penetration of the barrier.

Figure F.5 illustrates the construction of the double-stud exterior wall. It is shown as a bird's eye view. The order of construction is as follows:

1. First build the inner stud wall in the conventional manner (usually flat on the subfloor) using 2×4 studs 16″ on centers. Although this is the inner wall, it will carry the load of the ceiling and roof. The headers over the door and window openings should be sized as required by the code, i.e., 2×6, 2×8 and larger. Include both top plates.
2. Next place a 6-mil sheet of poly on top of the wall section. Allow about 18″ of excess poly around the perimeter for overlapping adjacent wall sections, the floor and ceiling.
3. Nail ¼″ to ½″ sheathing on top of the poly. Do not forget to make the cutouts in the sheathing for window and door openings

FIGURE F.5 Double-Stud Wall Assembly

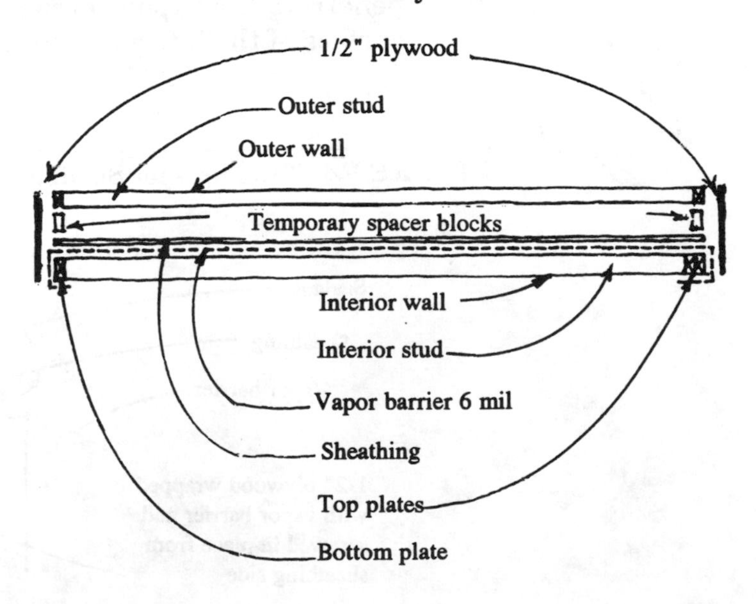

before placing on the wood wall frame. Be careful not to penetrate the poly. After installation the sheathing protects the poly.

4. Next, build the outer stud wall on top of the sheathing. Since this wall is non-load-bearing, it can be built of 2×3 studs placed 24″ on centers. Headers can be the same size.
5. Using temporary blocking, raise the outer wall to provide the desired space for the needed insulation.
6. Cut and apply ¼″ plywood plates across the bottom and top plates to connect the inner and outer walls.
7. Remove the temporary blocking and stuff the area outside the sheathing with insulation.
8. Tilt the wall section upright and install it in its proper position.
9. Tie the wall sections together with metal straps or 2×4 wood blocking over the top plate.

The wall can now be wired and insulated. Since the inner stud is on the inside of the vapor barrier wiring can be run through the studs without the need to penetrate the barrier.

Figure F.6 illustrates how vapor-barrier connections are made between various wall segments. The drawing is a view downward from the top of a double-stud wall.

FIGURE F.6 Double-Stud Wall Details

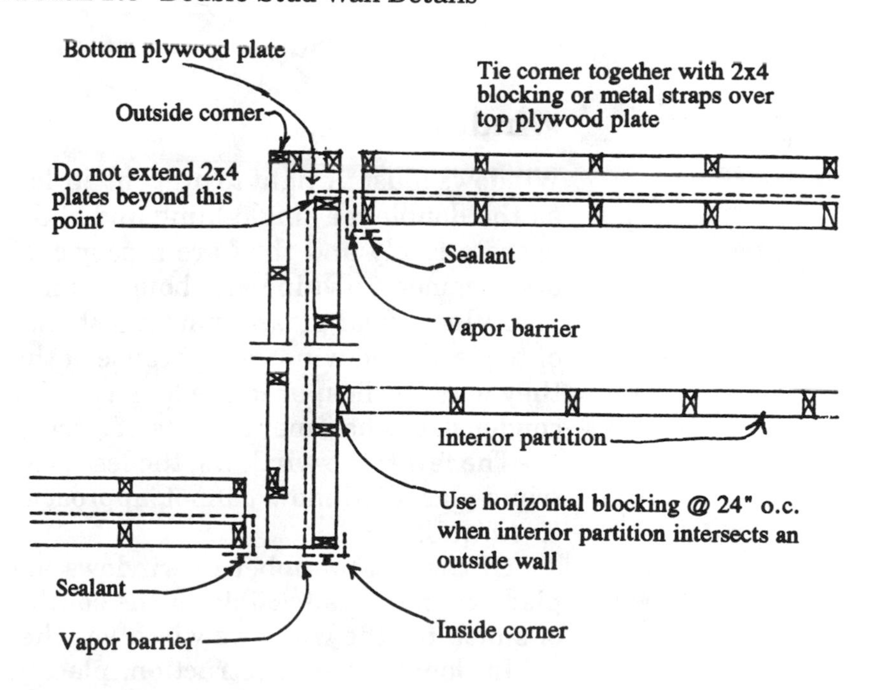

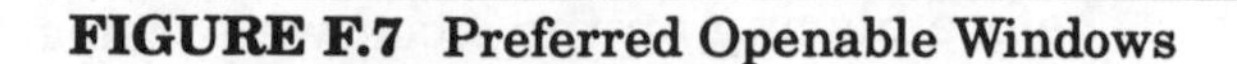

FIGURE F.7 Preferred Openable Windows

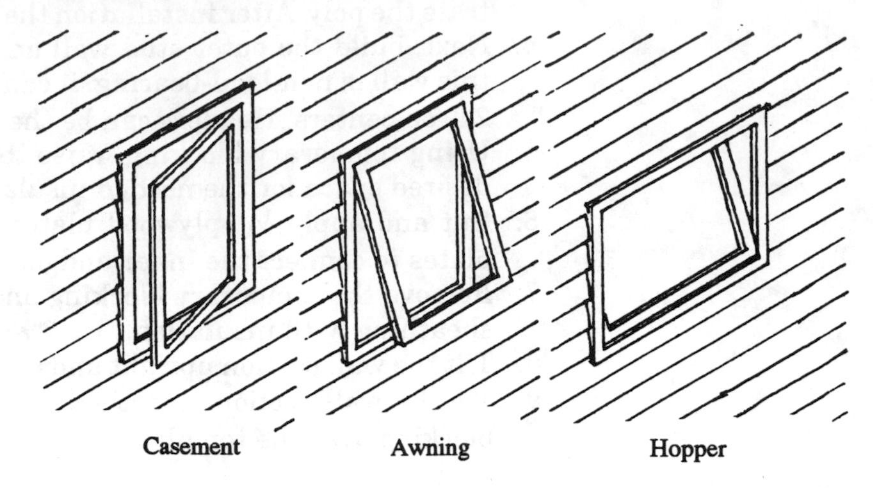

To prevent condensation within the walls, *there should be about twice as much insulation outside the vapor barrier as inside.*

For rooms with very high humidity content in normal usage, such as laundries or swimming pools, place the vapor barrier on the *inside* of the inner stud wall as shown in the single-stud construction in Figure F.1.

Windows

Windows must fit tight to prevent air leakage. Sliding windows such as the double- or single-hung and slider varieties must be loose to operate freely and thus are a poor choice. The preferred windows are casement, awning and hopper units (see Figure F.7).

At least dual-glazed windows should be used, but check the cost of triple-glazed windows. Because of their greater insulating value, they may be cheaper in the long run. In addition, less moisture will condense on the inner surface of a triple-glazed window.

The fewer the windows, the less loss of heating energy. Even the triple-glazed window cannot approach the insulation value of an R–42 wall.

Of the total number of windows and glass doors in the house, place as many as feasible in the south-facing wall. During periods of sunshine, the windows will allow the passage of solar heat.

In double-wall construction, place windows in the inside wall. They will be sheltered from the wind, and more air can circulate

FIGURE F.8A Vapor-Barrier Installation at Windows

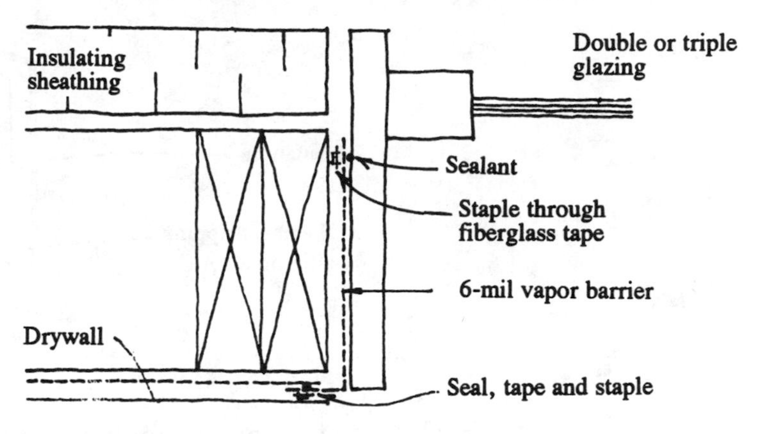

FIGURE F.8B Window Detail for Double-Stud Wall

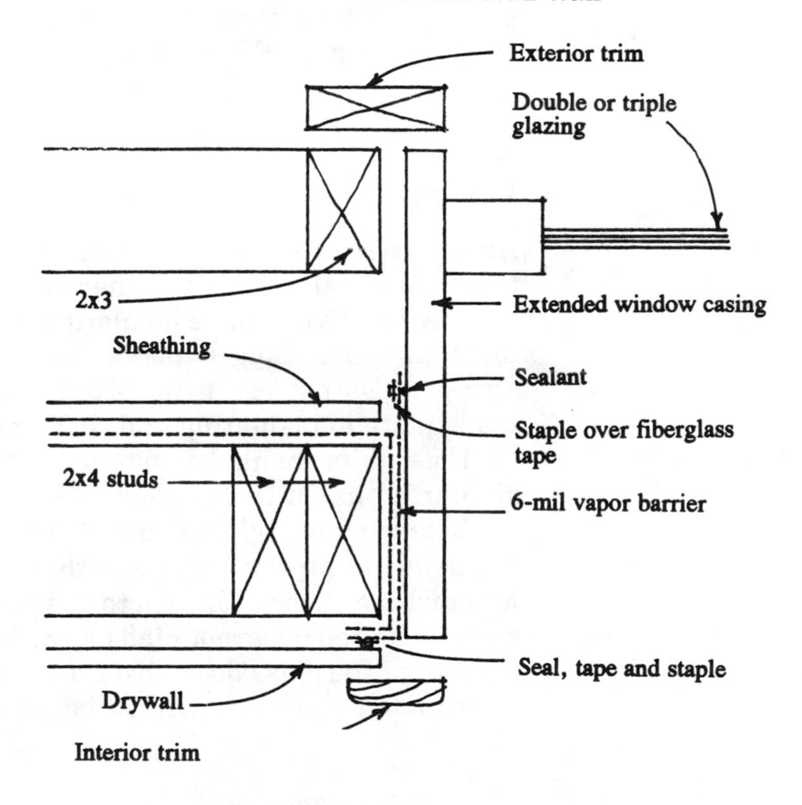

Note: Attach vapor barrier strip to window casing before inserting into rough opening. Allow sufficient material at corners to fold vapor barrier.

FIGURE F.9 Interior Shutter for Window

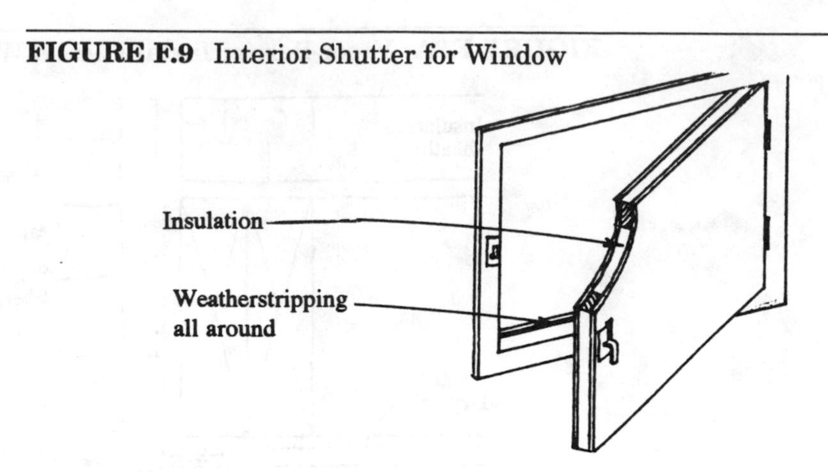

freely over the inside window surface, thus reducing moisture condensation.

Figures F.8A and F.8B illustrate the installation of the vapor barrier around a window in single- and double-wall construction.

To prevent undue heat loss, particularly passive solar heat, windows should have an interior shutter installed to prevent the passage of heat through the glass during periods of little or no sunshine. Figure F.9 illustrates a typical shutter.

Doors

As with windows, the best doors for the superinsulated house are those that seal well and in themselves are insulated, such as steel units with polyurethane insulating cores. Sliding doors usually leak more air than hinged doors that have been properly weatherstripped. Do not use hollow-core exterior doors. They are grossly inadequate in insulating and sealing capability.

Door units should be installed with a vapor barrier in a manner similar to that of the window illustrated in Figures F.8A and F.8B.

Like windows, glass doors also require an insulating shutter on the inside to prevent heat loss through the glass during periods of no sunshine. A good solution to this problem is the use of the modern steel insulated door mounted on tracks that permit sliding the door in front of the glass door. Order the doors *before* they are drilled by the supplier since you will not be using the normal door hardware.

Roof and Ceiling

The ceiling vapor barrier is stapled under the joists with seams at a joist to provide a firm base.

If it is practical in your house design, use prefabricated trusses for the roof framing. They will give you an uninterrupted span from exterior wall to exterior wall, and the complete ceiling vapor barrier can be installed easily. In this type of framing, the interior walls that are non-load-bearing are put up after the roof and ceiling structures are complete and the ceiling vapor barrier has been installed. The use of prefabricated trusses will reduce the usable attic space because of the cross-bracing inherent in this design.

In most cases, if your roof framing is the stick-built rafter-joist combination, you will usually have load-bearing interior walls that must be installed before the joists and rafters can be put up. In this case you will have to install the ceiling vapor barrier around these load-bearing interior walls. Figure F.10 illustrates a method of solving this problem for the house built over a crawl space. The bottom vapor barrier is not needed for a house with a basement or a concrete first floor.

Conventional roof ceiling joist construction does not usually provide enough vertical height in the attic space near the eaves for the necessary amount of insulation. For that reason, make sure that the bottom of the soffit framing and the bottom of the ceiling joist are at the same level. This will require the raising of the roof rafter by installing a plywood gusset plate, as shown in Figure F.11. This requirement applies to both the stick-built roof and manufactured trusses.

FIGURE F.10 Vapor-Barrier Installation for Partition Wall

THE SUPERINSULATED HOUSE

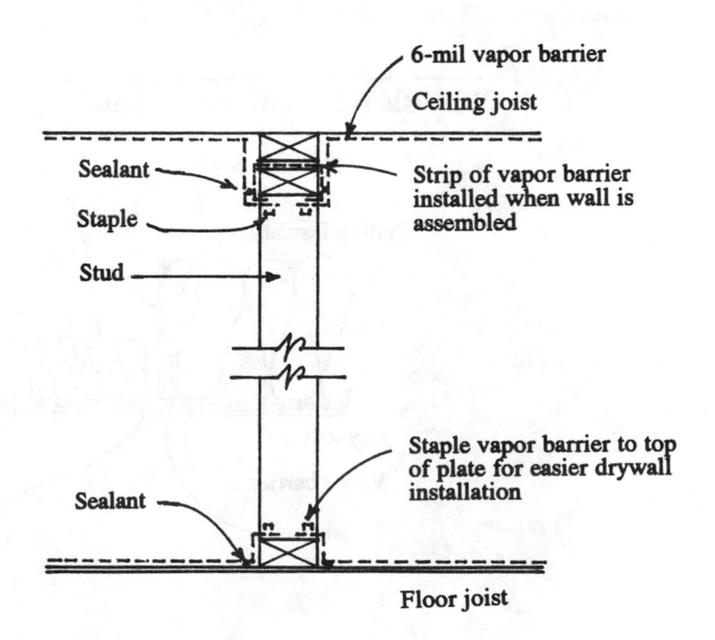

FIGURE F.11 Roof Construction

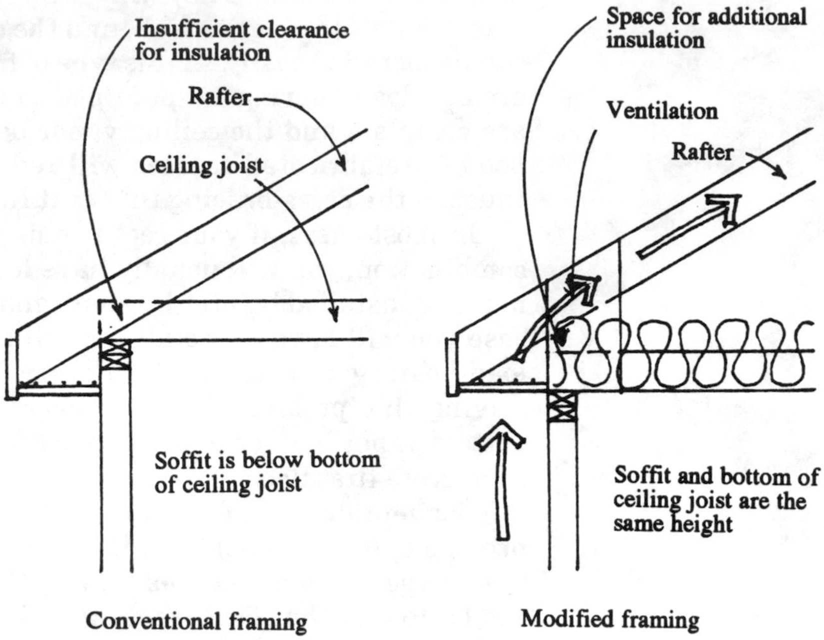

Insufficient clearance
for insulation
Rafter
Ceiling joist
Soffit is below bottom
of ceiling joist
Conventional framing
Space for additional
insulation
Ventilation
Rafter
Soffit and bottom of
ceiling joist are the
same height
Modified framing

Roof and Ceiling Penetrations

Entry into the attic space is a necessity. Even if the attic is not used for storage, an entry is needed to maintain the structure. If possible install this entry from an unheated portion of the house such as the garage. If it must be installed within the heated area of the house,

FIGURE F.12 Attic Hatch Detail

Plywood box
Burlap cover
Insulation
Ceiling insulation
Drywall
Vapor barrier
Sealant
Weatherstripping around
perimeter

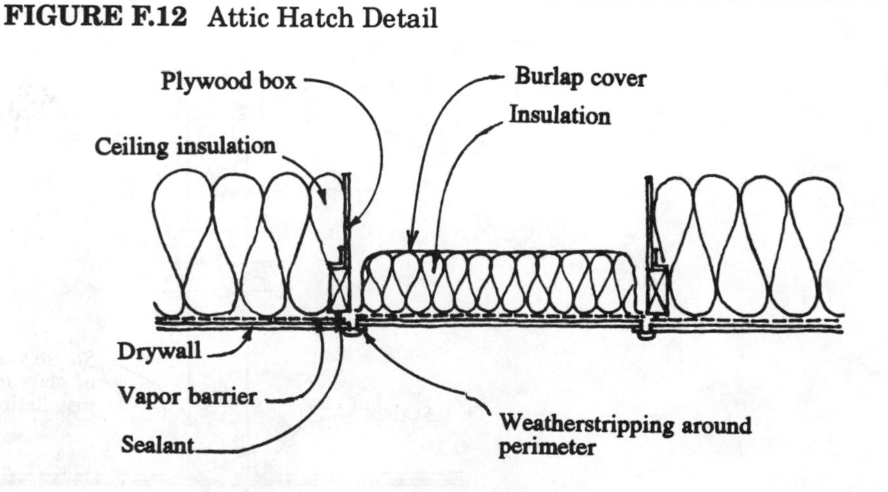

FIGURE F.13 Vapor-Barrier Installation at Plumbing Vent Stack

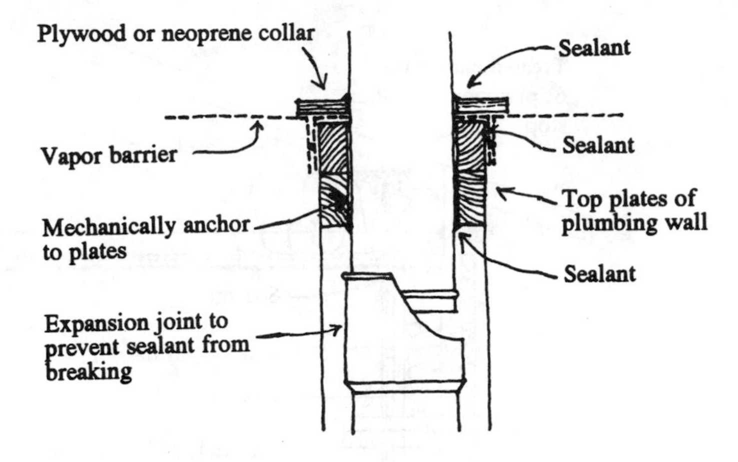

Figure F.12 illustrates a method of doing so and sealing the area when not in use.

Figure F.13 shows proper sealing around a plumbing vent stack. Figure F.14 illustrates a method of sealing around a chimney installation. The dual-walled insulated metal chimney is the preferred type to use.

Typical Construction Methods

Figure F.15 illustrates a method of building a house using the all-wood foundation system. One advantage of this system is that it

FIGURE F.14 Chimney Installation Detail

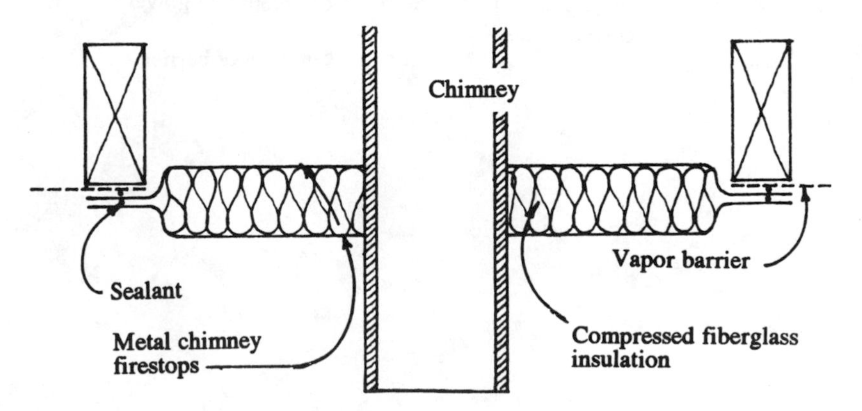

FIGURE F.15 Wall Section

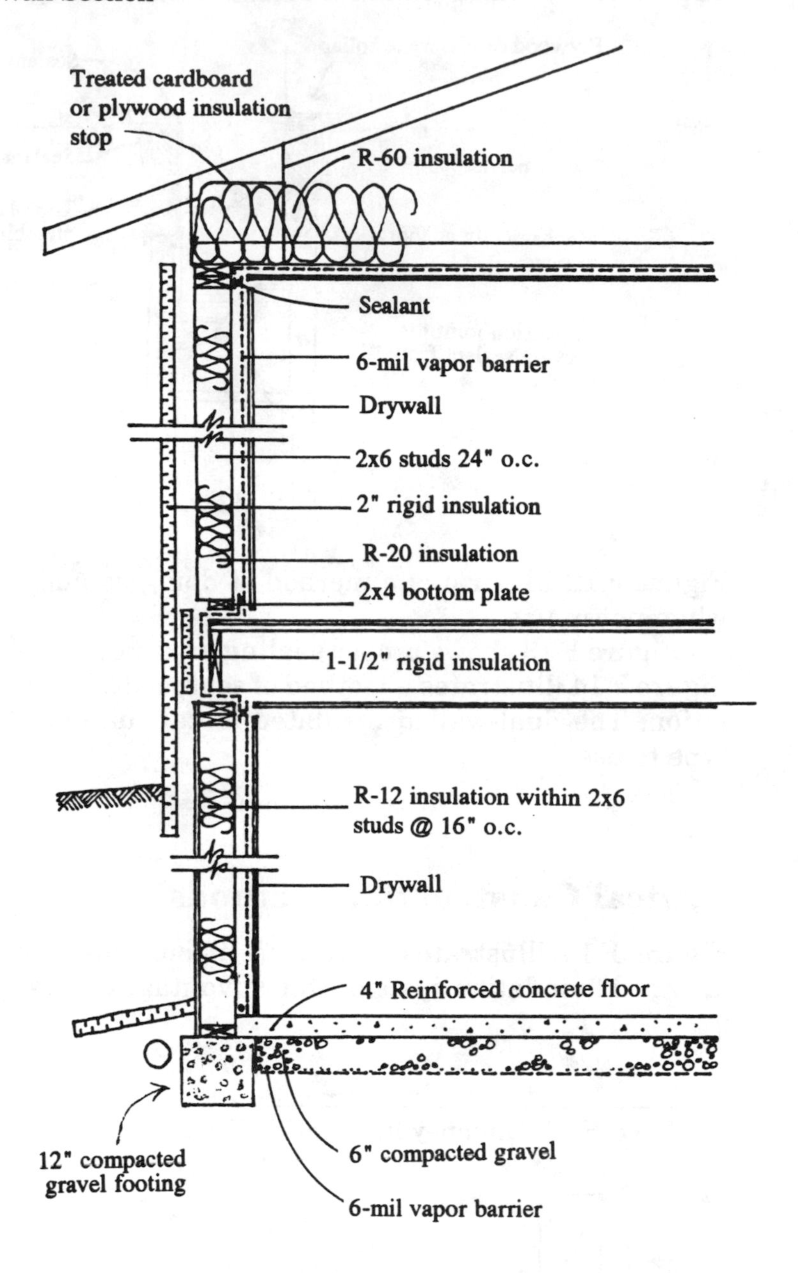

FIGURE F.16 Wall Section: Concrete Wall and Footing

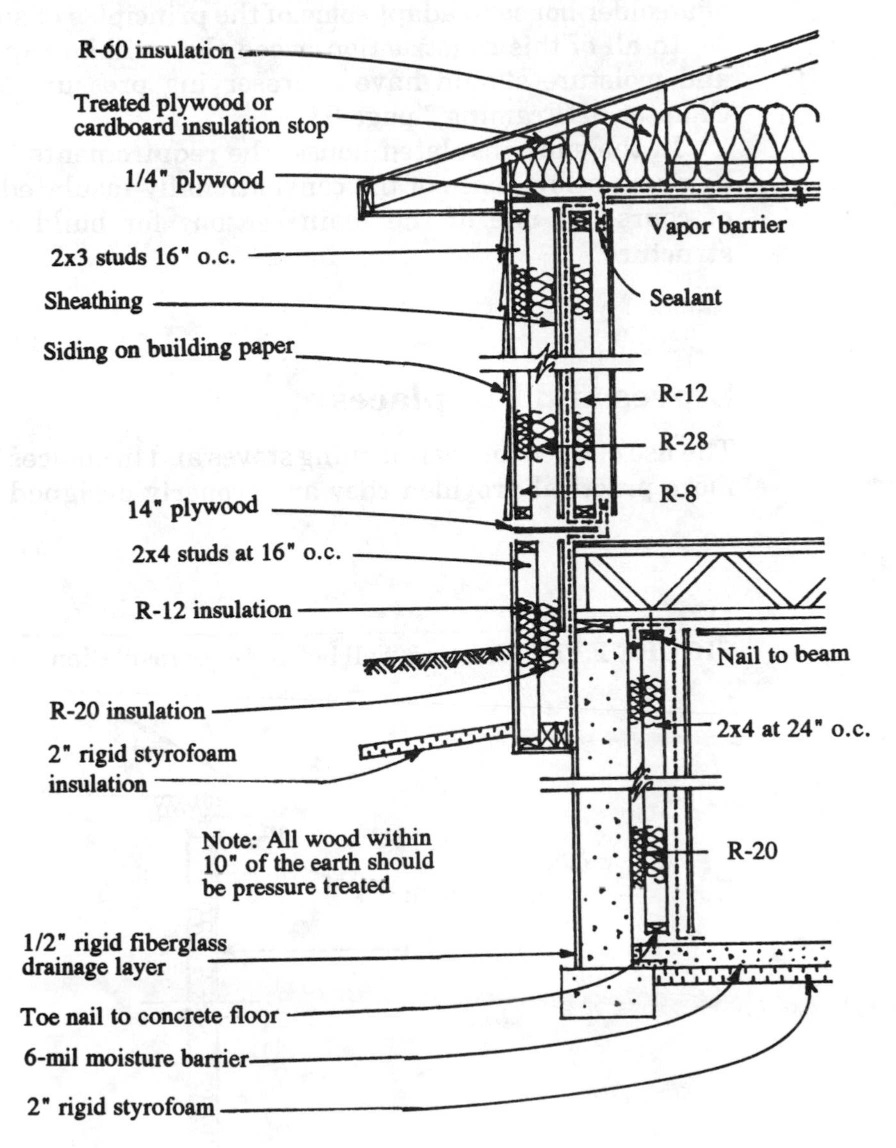

can be built in climates that are too cold to pour concrete or lay block and brick. Note the use of cardboard as an insulation stop to ensure that the insulation does not fill the air space necessary for proper attic ventilation.

Figure F.16 illustrates the techniques used for a concrete foundation and basement. Note the use of the wood truss system for floor joists. The open spaces provide room for the easy installation of heating ducts, plumbing and electrical cables.

Figures F.17A and F.17B illustrate modifications to the structure of an older house to adapt some of the principles of superinsulation.

In all of this construction, wood that will be exposed to the air and moisture should have a preserving pressure treatment. See Chapter 6, "Framing," page 63.

In the superinsulated house, the requirements for heating will be much below those of the conventionally insulated house, which, of course, is one of the main reasons for building this type of structure.

Stoves and Fireplaces

The use of wood- or coal-burning stoves and fireplaces becomes much more practical provided they are properly designed and installed.

FIGURE F.17A Existing Wall before Superinsulation

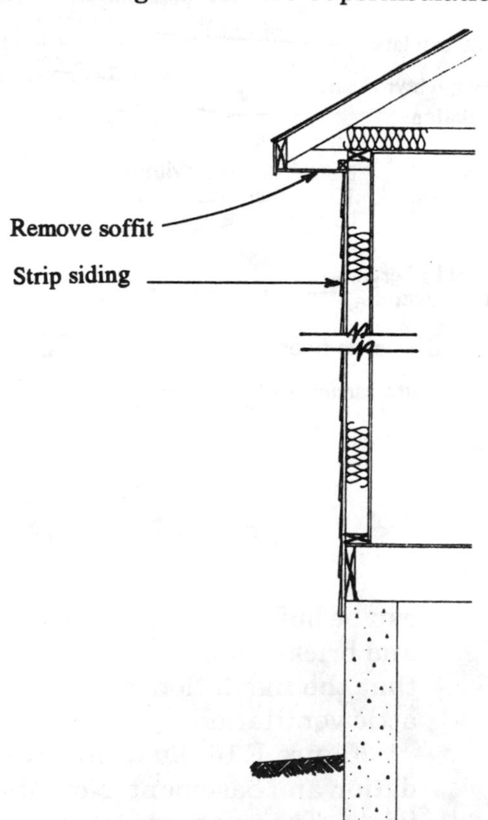

FIGURE F.17B House after Superinsulation Heating Systems

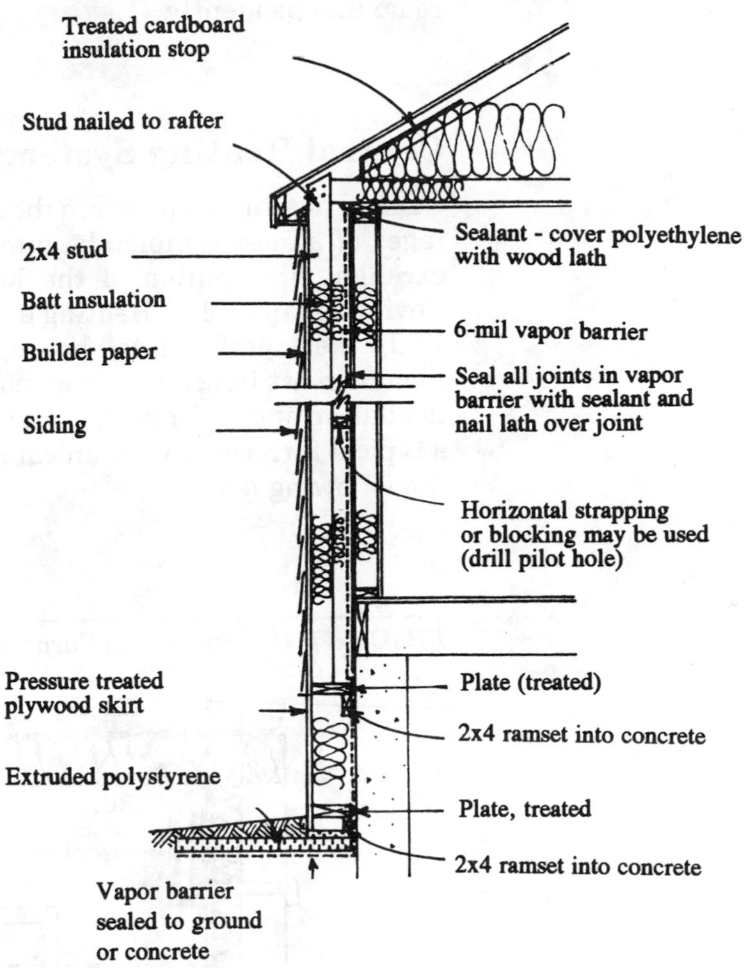

An efficient fireplace (metal) is illustrated in Chapter 10, "Masonry and Concrete," pages 114 and 115. A stove should have similar characteristics.

Electric Baseboard Heat

This type of heating system should also be considered. It is more effective and efficient in a superinsulated house than it would be in a house with conventional insulation and anti-air-infiltration measures. It is much less costly to install than forced-warm-air systems, and it offers the flexibility of regulating the temperature of each room independently.

Central Heating Systems

Central heating systems are the most popular. They have the advantages of almost completely automatic operation, and they provide excellent distribution of the heated air through a duct system. Review Chapter 14, "Heating and Cooling."

In the superinsulated house, proper construction of the furnace room is most important so as not to reduce the effectiveness of the insulation and sealing of the rest of the house. Figure F.18 illustrates a typical furnace or mechanical room located in the basement. Note the following features:

FIGURE F.18 Isolation of Furnace Room

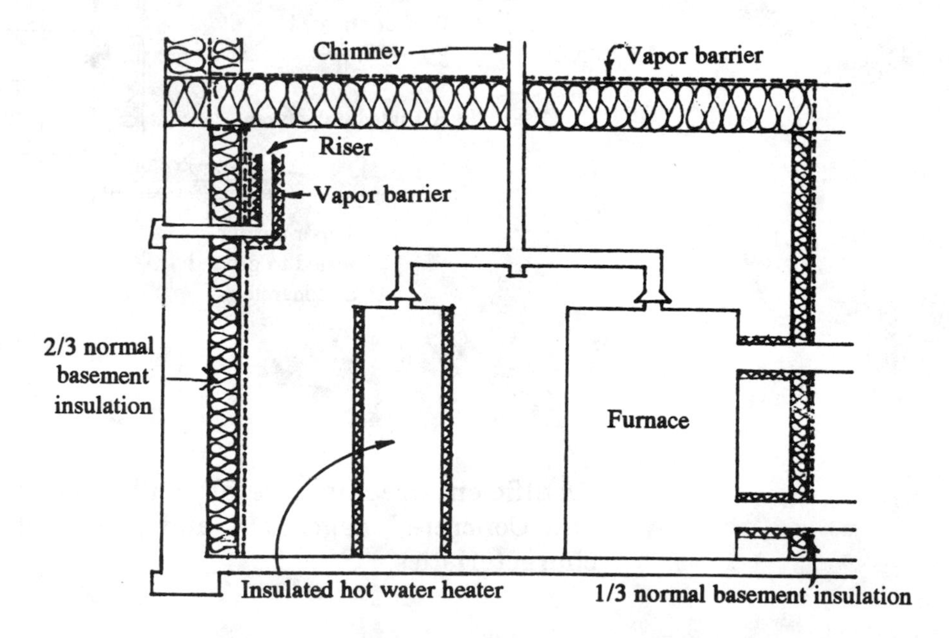

1. The room is sealed from the rest of the house with a poly vapor barrier. This is necessary because outside air to support combustion moves freely into the room through the riser vent.
2. The purpose of splitting the normal basement insulation is to permit enough heat to seep through the interior wall to prevent the freezing of the pipes of the water heater.

If your house has no basement, a furnace room can be built on the first floor.

In an energy-efficient house, very little heat is needed, so buy a small furnace. Get one with electronic ignition and a positive chimney damper.

Passive Solar Heat

In a properly designed superinsulated house, passive solar heating can provide 30% or more of the heat needed instead of the more usual 5% to 10%. Thus, any additional cost in construction to provide an effective passive solar system is economical in most cases. Review Chapter 13, "Solar Heating."

Water Heating

In many cases, the superinsulated house consumes more fuel to make hot water than it does to heat the house. Figure F.19 illustrates a two-tank system, which will reduce the cost of heating hot water.

Water entering the house during the winter months will probably have a temperature around 40° F. By installing the second uninsulated tank and piping the water into this tank first, it is heated up to about the temperature of the air in the mechanical room. This practice substantially reduces the load of the actual water heater. While this system is taking heat from the house to preheat the water, there is usually an excess of heat in a superinsulated house for most of the cold months.

Additional savings can be made in a large house by installing two or more smaller water heaters close to the fixtures that will use the water. This design reduces the amount of water needed to fill the supply pipe before it comes out at the fixture. Since the hot water remaining in the pipe will soon lose its heat, the shorter this pipe, the less the loss of energy.

FIGURE F.19 Technique for Preheating Hot Water and Insulating Hot-Water Tank

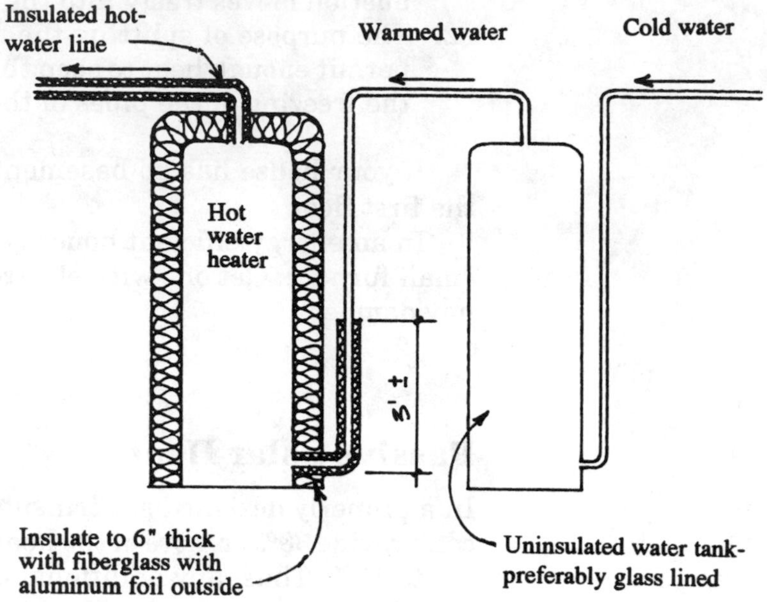

Ventilation

Unlike the conventional house, the superinsulated house will have ventilation and dehumidification problems. Since the house is tightly sealed, the moisture generated by normal living activities such as showering and clothes washing will remain in the house, as will odors from cooking and other functions. This lack of fresh air will make living uncomfortable and must be corrected.

You can reduce this excess moisture by using a dehumidifier. By far the best solution, however, is the ventilation of the house by a mechanical heat exchanger.

The Air-to-Air Heat Exchanger

The heat exchanger, Figure F.20, is a mechanical device that brings fresh air into the house while passing the stale air out and exchanging the heat from the stale air to the fresh air in the process. The fresh-air temperature, therefore, should be close to the temperature of the room. Heat exchangers are about 75% to 85% efficient.

The heat exchanger should be installed in the basement or in the mechanical room along with the furnace and hot water heater. It should *not* be installed in an unheated crawl space because the

FIGURE F.20 Schematic of Air-to-Air Heat Exchanger

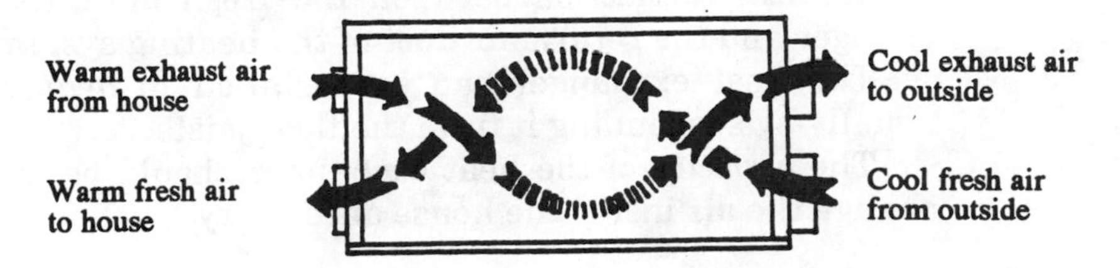

moisture from the stale air will freeze and the exchanger will not operate properly.

Figure F.21 illustrates the installation of a heat exchanger system in a house that does not have a forced-air heating system. Note that to operate efficiently, the exchanger must have its own duct system, which extracts the stale air from rooms that usually generate the most moisture (bathrooms, laundry and kitchen) and brings fresh air into the other parts of the house.

In houses with central forced-warm-air heating systems, the fresh air supplied by the exchanger should be ducted to a room where it can mix with the room air and then be picked up by the return air

FIGURE F.21 Fully Ducted Heat Recovery System

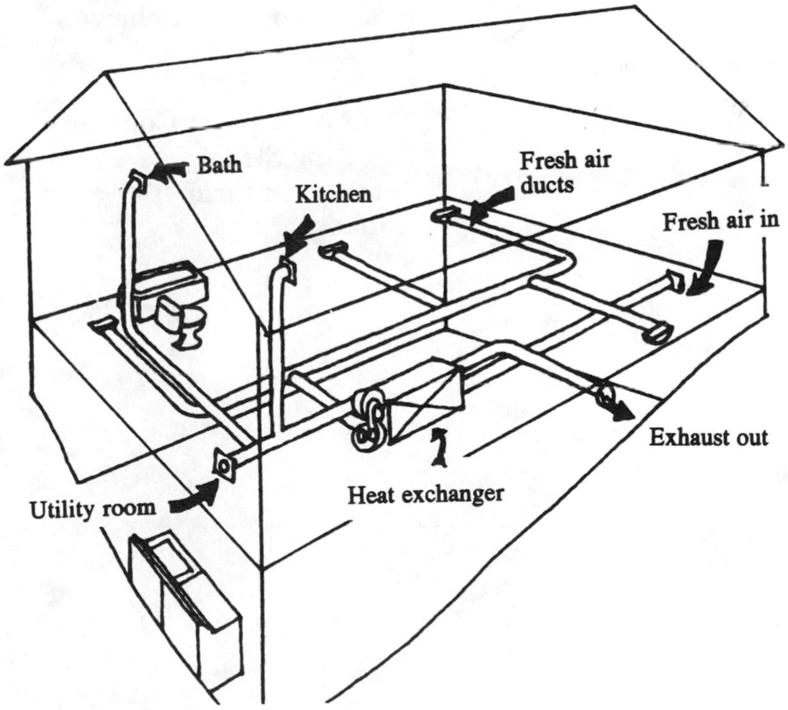

ducts of the central heating system. There should *not* be any closed mechanical connection between the fresh air duct from the exchanger and the return air duct of the heating system.

The heat exchanger can be installed in *heated* crawl space usually by suspending it from the floor joists.

The capacity of the heat exchanger should be large enough to change the air inside the house once every two hours.

Acknowledgments

The contents of this appendix are based largely on information provided by the following sources:

Energy Conservation Branch
Saskatchewan Energy and Mines
1914 Hamilton Street Regina, Canada

Small Homes Council—Building Research Council
University of Illinois at Urbana-Champaign
Champaign, Illinois

Conservation Energy Systems, Inc.
Box 8280
Saskatoon, Saskatchewan
Canada

The Air Changer Company
334 King Street
Toronto, Ontario
Canada

Glossary

active solar heat Heating system using heat from the sun as its sole source of energy and using mechanical means for distributing the heat throughout the house.

air infiltration Movement of air into and out of the heated space of a house through cracks in floors, walls and ceilings.

amortization The repayment of the principal of a mortgage.

ampere An expression of the amount of electricity moving through a circuit. Also, an expression of the capacity to carry electricity for a cable or wire.

anchor bolts Large (usually ½″ diameter and 12″ to 16″ long) bolts set in the foundation wall or slab (when foundation is an integrated slab) that anchor the sole plate to the foundation.

angle iron Steel bars shaped like an "L" with variable length. Used as lintels to support brick veneer over window and door openings.

apron Piece of wood interior trim under the window stool (sill).

awning window Window that pivots on a horizontal hinge at the top.

backfill Material, usually dirt, sand or gravel used to change the level of the ground for specific purposes such as to fill in the voids around the house between the excavation for the basement and the basement wall.

back prime Primer paint applied to the back side (nonexposed side) of exterior wood to prevent warping of the wood.

balloon framing Framing system in a two-story house in which the studs of the outer walls are one continuous piece from the first floor to the roof system.

baseboard Interior trim, usually wood, laid at the base of the wall to cover the gap at the wall-floor seam.

batt Insulation in the form of rolls in various widths, lengths and thicknesses.

batten The outer board of board-and-batten siding.

batter boards Wood boards that hold the strings used to mark the outer limits of the foundation. They are removed after the foundation is complete.

bay window A window composed of three or more flat sections that juts out from the exterior wall.

beam Also called a girder. A structural member, usually wood or steel, used to hold heavy loads over relatively long spans.

bid An offer from a contractor or supplier to provide labor and/or material for a certain price.

bifold door A type of folding door normally used on closets.

blocking A method of laying material between joists and rafters to prevent twisting and to increase the strength of the framing.

blueprints The drawings indicating the plans and elevations for a structure.

board foot Measurement unit for lumber. One board foot is the equivalent of a board 1″ thick (nominal), 12″ wide (nominal) and 12″ long (actual).

bow window A curved window composed of several sections that juts out from the exterior wall.

bridging A method of installing specially cut wood or steel bracing to prevent twisting of joists and rafters and to strengthen the overall framing.

brick molding Molding used as exterior trim around doors and windows to finish the joint of the window or door and the siding, both brick and wood.

brick ties Galvanized corrugated steel straps about 1″ wide and 4″ long used by the mason to tie the brick to the wood framing.

brick veneer A single thickness of brick laid on the outside of a masonry or wood-framed wall as the exterior finish.

building code A set of building standards adopted by a local governing body to establish minimums for building houses.

building paper A heavy black paper used in the building trade to prevent moisture seepage.

building setback lines Lines that parallel the front, back and sides of the lot to establish the area in which the house must be built.

built-up roof A form of roof finish consisting of several layers of roofing paper with tar in between. Usually topped with gravel. This type of roofing is used primarily where the roof pitch is below $3/12$.

cantilever A part of the house structure that extends out from the exterior wall such as a bay or bow window, or, in some colonial plans, the second floor projects out over the first floor 2′ or so.

casement window A window that swings on a vertical axis like a door.

casing Wood interior trim applied around windows and doors.

cathedral ceiling Framing structure that eliminates the ceiling joists and exposes the underside of the roof to the living area.

caulking The process of applying latex or similar material to seal exposed wood, aluminum and vinyl joints and seams.

cement A powdered binding element that is mixed with water, sand and aggregate (stone) to form concrete.

chair rail Wood interior trim applied to the wall at approximately chair-back height to protect the wall and improve design.

chase Usually refers to a "duct chase," part of the house framing in which heating and cooling ducts are installed. Separated from the room by drywall, plaster or paneling.

chimney cap A metal device applied over the top of the chimney to prevent rain and snow from falling down the flue.

circuit breaker An electrical safety device located in the service panel that automatically opens the circuit (stops the flow of electricity) when the capacity of the circuit is exceeded. Can be easily reset. Replaces the older fuse system.

clearing The process of removing trees and other growth from a lot so that construction can begin.

collar beam A brace between rafters in roof construction.

colonial house By common usage, a style denoting a two-story house in which both floors have about the same square footage. There is a true architectural colonial design, however, that has distinctive features not included in the majority of two-story house designs.

concrete A mixture of cement, water, sand and an aggregate (stone) that hardens into a solid building material.

concrete block Building blocks formed of concrete with a height of 8″, length of 16″ and widths varying from 4″ to 12″. Used primarily for the construction of foundations and walls.

corbel Extending bricks out of pattern to provide a base for material installed on top, usually concrete.

corner board Vertical pieces of wood installed at the corners of the house to trim lap and other wood siding.

crawl space The space between the grade of the lot and the floor joists of houses with wood first floors.

cricket A form of roof flashing applied between the upper side of a chimney and the roof, shaped to prevent water settling at this point.

crown molding Wood interior trim installed at the junction of the ceiling and wall.

culvert A concrete or metal pipe that allows water to flow in a drain ditch underneath the driveway.

cupping A form of distortion in thin lumber that results in the lumber curling around the long axis.

curing A process used to attain maximum strength in freshly poured and troweled slab concrete by keeping it moist over a period of time (at least several days).

diffuser A metal cover on the room outlet of the air supply that directs the flow of air.

direct gain The acquisition of passive solar heat by collection through a window and storage in a wall or floor.

direct hire Employment of labor directly in which the employer is responsible for payroll deductions, worker's compensation and so forth.

dormer A window whose framing protrudes through the roof.

double-hung window Window design in which both the upper and lower sashes move vertically within slots in the frame.

door stop Wood trim around the inside of the door frame (jamb) against which the door rests when closed.

dropped ceiling A ceiling, consisting of a light steel frame with rectangular composition board fillers, hung below the standard house ceiling.

drywall An interior wall finish material consisting of large boards of gypsum with paper finish.

dual glazing Also called insulating glass. Double panes of glass with air space between.

duct A large conductor made of fiber glass or galvanized sheet metal that carries heated or cooled air throughout the house.

earthstone Hand-molded clay tile floor covering.

elastomeric Synthetic rubberlike material used to waterproof wood and masonry walls above and below grade and flat built-up roofs. Has a high degree of elasticity even in very cold weather and resists the effects of the sun to a greater degree than the tar and paper built-up roof.

elevation A drawing of the vertical sides of a house. Also, the relative level of the ground at any particular point.

envelope house A house designed so that both cool and warm air are circulated around the house from crawl space to side wall to ceiling to side wall to crawl space.

equity The value of a house and lot (or any real estate) beyond the total amount owed on it in mortgages, liens and other debt instruments.

exterior trim Wood or aluminum trim material applied to the exterior of a house.

facia A trim board that runs parallel to the eaves of a roof.

fill Material, usually dirt, sand or gravel, applied to raise the elevation of the ground.

flashing Metal or vinyl material used to divert water from junctions in the roof.

flint paper A special paper sold by building supply stores for use in covering finished floors to protect them until the construction is finished.

flitch plate A plate of steel or plywood bolted or nailed between 2″ lumber to form a strong beam for relatively long spans.

folding door A door made of wood plastic or other material that folds within the door frame.

footing The concrete (or gravel in the case of the all-wood foundation) upon which the foundation rests.

form Material, usually wood, built to hold concrete in place until solidified.

forced warm-air heating A heating system based on the principle of forcing heated air through a duct system.

friction fit Usually refers to insulation batts that do not have an integrated vapor barrier. Held in place in the walls by the friction of the insulation material against the studs.

frieze board Wood trim board applied at the juncture of the siding and the soffit.

grade stake A wood or metal peg placed so that the top of the peg indicates the final level of fill dirt, concrete or other material used to change the grade or elevation.

grading The task of using machinery or hand tools to change the level of the ground to conform to the construction requirements, such as the floor of a garage, a patio and so forth.

header A wood beam placed in the framing to span the opening of a door, window, pull-down stair, skylight, etc.

headwall A masonry wall built at each end of a culvert to prevent erosion of the driveway fill and to improve appearance.

hearth The part of the fireplace, usually built of brick, slate or tile, extending into the room from the fireplace opening.

heat pump A heating system based on transferring heat rather than manufacturing it.

house layout The establishment of the vertical and horizontal location of the foundation of a house by setting up the batterboard system.

HVAC An acronym for the heating, ventilating and air-conditioning system.

index A term used in mortgages, particularly those with variable rates, which establishes the base from which the interest rate of the mortgage may vary. If the index used to compute the mortgage increases, the interest rate of the mortgage also increases. Examples of such indexes are the Federal Home Loan Bank Board's national average mortgage rate, the U.S. Treasury Bill rate and the prime rate.

insulation The material used to establish a barrier to the movement of heat through the outer skin of a house living area.

insulating glass Two or more glass panes in a frame with insulating air space in between.

job built Items such as bookcases that are built on the job site rather than in a factory or shop.

joist The load-bearing beam in a floor or ceiling.

let-in Relates to structural members that are cut into other members such as corner bracings.

lintel Steel or concrete beams that span openings in the masonry wall of a house over doors and windows.

masonry Work consisting of the use of brick, stone or block.

mil A measurement, $1/1{,}000'$, used to determine the size of wire and the thickness of a sheet of polyethylene.

mortar joint The joint between bricks, stone, block and tile and the finish treatment of the mortar.

mud A slang term referring to the application of tile, brick, stone or similar material into a bed of mortar, usually in floor or wall finishing.

mullion The vertical member between the sections of a double, triple or larger window.

muntins The vertical and horizontal wood, plastic or metal members between the panes of glass in a window or door. For most insulating windows or doors, the muntin is a decorative device placed between the glass sheets covering the entire sash and does not separate panes of glass.

o.c. Means "on centers" and refers to the distance between the centers of parallel structural members such as studs—16″ o.c.

panelized Refers to a method of home construction in which sections of the wall are factory built in panels, transported to the site and assembled by the framing crew.

parging The placement of a coating of mortar over a masonry wall, generally for waterproofing purposes.

passive solar heat The use of heat from the sun to warm a house or its parts without any machinery except small fans to assist in circulation.

permit An authorization by the local government to perform certain construction.

picture frame A method of trimming the interior of a window in which the stool and apron at the bottom of the window are replaced by the same type and size of trim used at the top and sides of the window, giving the appearance of a picture frame.

pier A masonry or wood vertical member that supports floor beams or girders. Principally used in crawl-space construction.

pilaster The thickening of a masonry, concrete block, brick or concrete wall at selected points to increase its strength.

plan A drawing, usually supplemented by written specifications that indicates the details of construction of a house. Also, drawings that show horizontal views of the house, in contrast to elevations, which show the vertical views of the house.

plank and beam framing A method of framing in which heavier timbers with greater spans are used.

plastic housewrap A thin, about 6 mil, plastic roll sheet material used to wrap the exterior walls of the house over the sheathing. It prevents air from moving through the wall and reduces the cost of energy for cooling and heating.

plastic laminate Layered plastic material manufactured in a variety of colors and designs used for countertops and similar items in house construction.

plaster veneer An interior wall finish in which one or two thin coats of plaster are applied to a plaster lath base.

platform framing A method of framing in which each floor has its own set of studs.

plate A horizontal structural member in the wall. There are three types in platform framing: sill, sole and top plates.

pocket door A door that slides into a slot in the adjacent wall.

polyethylene A plastic sheet material used largely for vapor barriers in walls and on the ground in crawl spaces.

prehung door A door that is delivered to the job with the frame and hinges attached. May also have the casing partially installed.

pulldown stairs Stairs used to provide access to the attic that fold up into the ceiling and attic.

R–value A number indicating the relative efficiency of a material in insulating from heat loss. The higher the R, the more efficient the material for insulation purposes.

radiant heating A heating system in which the heat is transferred principally by movement through the air as a wave similar to light.

rafter A structural member of the roof.

rancher A house design on one floor.

rake molding Trim molding applied on top of the rake board.

reinforcing bar (rebar) Steel bars with diameters of ¼″ and greater used to increase the strength of masonry and concrete.

retaining wall A wall designed to prevent erosion and collapse of a dirt bank.

ridge board A wood structural member at the peak of the roof against which the rafters are nailed.

row lock A method of laying brick in a sloped horizontal position, usually used on the exterior under windows and doors.

rough opening The size of the opening in the framing needed to accept a window, door or other item that is to be installed within the framing.

salt-treated lumber Lumber into which a salt solution has been forced by high pressure to increase its ability to withstand moisture and insects.

section A type of drawing that represents the details of part of the structure as though that part had been sliced vertically by a large knife.

septic system A sewage disposal system that releases sewage to be absorbed into the ground on which the house is built.

shake A wood shingle, usually cedar, that has been split by hand rather than machine.

sheathing Material, usually 4′ × 8′ or 2′ × 8′, applied to exterior walls and roofs directly to the studs and rafters. May be plywood, impregnated composition board or rigid-insulation material.

shingle molding Molding applied over the rake board just under the shingles.

shoe molding The interior wood trim that is installed to fill the gap between the baseboard and the finished floor. Generally not used with carpeting.

siding Material applied to the outside of the exterior walls.

sill cock Exterior water faucet.

sill plate The wood plate on top of the foundation that supports the floor joists.

sizing Adhesive applied over bare plaster and drywall to provide better adhesion for wallpaper and other wallcovering.

skylight A window installed in the roof.

soffit Underside of the roof overhang.

soil pipe Waste (drain) pipe from a toilet.

specifications (specs) Written details of the overall plan that support and augment the blueprints.

SPF Framing lumber consisting of a mix of spruce, western pine and fir.

splash block A preformed block of concrete or other material placed at the end of a downspout to prevent erosion of the yard.

spoil Material, usually dirt, removed from an excavation.

square A term pertaining to the area of the roof meaning 100 square feet.

stick built A slang term for the method of framing in which the structure is job built from uncut lumber.

stool Another name for the interior window sill.

stucco A cement type of siding that is applied over a masonry wall or a wood frame (on metal mesh lathing).

SYP southern yellow pine.

take-off The computation of the materials required for construction of a house such as lumber take-off or masonry take-off.

tempered glass Glass that has been specially treated to minimize the danger of cutting when broken.

top soil Dirt, usually lying on top of the lot, that is very fertile and suitable for supporting the growth of grass and shrubs.

Trombe wall A masonry storage wall used in passive solar heating systems.

trowel finish A very smooth concrete finish suitable for interior floors and garages.

truss A plant built structure used in roof and floor construction.

U–value A numerical value assigned to different materials to indicate their relative efficiency in insulation. U is the reciprocal of R; hence the higher the U the *lower* the ability of the material to insulate.

vapor barrier A material, such as polyurethane, applied to walls, ceilings and floors on the inside of the living space to prevent moisture from moving into the insulation of the house and making it less effective.

vent Metal or wood pieces that permit the ventilation of crawl spaces. Metal vents can be closed to reduce air flow during the winter.

volt An indication of the pressure pushing electricity through a conductor.

wainscot Wood, ceramic tile or other material applied to an interior wall to a height of about 3′ to 4′.

wallcovering A general term for material used to decorate walls, such as wallpaper, fabrics, vinyls and precovered paneling.

waste pipe Pipe that carries the waste from the fixture to the sewer.

water hammer Noise in pipes caused by the sudden stoppage of fast-running water when the valve is turned off. It may also be caused by turning off one valve when another valve in the house is partially open.

watt The power required to operate electric equipment. It is the product of the amperes times the voltage.

whole house exhaust fan A large fan installed in the ceiling or attic that will evacuate the air from the whole house in a reasonably short time.

Index